SAFETEA-LU 1808: CMAQ
Evaluation and Assessment

U.S. Department of Transportation
Federal Highway Administration

FHWA-HEP-09-026

NOTICE

This document is disseminated under the sponsorship of the U.S. Department of Transportation in the interest of information exchange. The U.S. Government assumes no liability for the use of the information contained in this document.

The U.S. Government does not endorse products or manufacturers. Trademarks or manufacturers' names appear in this report only because they are considered essential to the objective of the document.

QUALITY ASSURANCE STATEMENT

The Federal Highway Administration (FHWA) provides high-quality information to serve Government, industry, and the public in a manner that promotes public understanding. Standards and policies are used to ensure and maximize the quality, objectivity, utility, and integrity of its information. FHWA periodically reviews quality issues and adjusts its programs and processes to ensure continuous quality improvement.

REPORT DOCUMENTATION PAGE			Form Approved OMB No. 0704-0188
Public reporting burden for this collection of information is estimated to average 1 hour per response, including the time for reviewing instructions, searching existing data sources, gathering and maintaining the data needed, and completing and reviewing the collection of information. Send comments regarding this burden estimate or any other aspect of this collection of information, including suggestions for reducing this burden, to Washington Headquarters Services, Directorate for Information Operations and Reports, 1215 Jefferson Davis Highway, Suite 1204, Arlington, VA 22202-4302, and to the Office of Management and Budget, Paperwork Reduction Project (0704-0188), Washington, DC 20503.			
1. AGENCY USE ONLY (Leave blank)	2. REPORT DATE July 2009	3. REPORT TYPE AND DATES COVERED CMAQ Evaluation and Review 2008	
4. TITLE AND SUBTITLE SAFETEA-LU 1808: Congestion Mitigation and Air Quality Improvement Program Evaluation and Assessment: Phase II Final Report		5. FUNDING NUMBERS	
6. AUTHOR(S) Terrance Regan, Elizabeth Murphy, Mary Beth Hines			
7. PERFORMING ORGANIZATION NAME(S) AND ADDRESS(ES) U.S. Department of Transportation Volpe National Transportation Systems Center 55 Broadway Cambridge, MA 02142		8. PERFORMING ORGANIZATION REPORT NUMBER FHWA-HEP-09-026	
9. SPONSORING/MONITORING AGENCY NAME(S) AND ADDRESS(ES) Office of Natural and Human Environment Federal Highway Administration 1200 New Jersey Ave., SE Washington, DC 20590		10. SPONSORING/MONITORING AGENCY REPORT NUMBER HEPN-1	
11. SUPPLEMENTARY NOTES This report was overseen by a review panel composed of representatives of the Federal Highway Administration, Federal Transit Administration, and U.S. Environmental Protection Agency.			
12a. DISTRIBUTION/AVAILABILITY STATEMENT No restrictions. This document is available to the public electronically through the Federal Highway Administration Office of Natural and Human Environment, Washington, DC 20590, at http://www.fhwa.dot.gov/environment/cmap=qpgs/index.htm.		12b. DISTRIBUTION CODE	
13. ABSTRACT (Maximum 200 words) In SAFETEA-LU Section 1808, Congress required the U.S. Department of Transportation, in consultation with the Environmental Protection Agency (EPA), to evaluate and assess the direct and indirect impacts of CMAQ-funded projects on air quality and congestion levels to ensure the program's effective implementation. Phase II of the *CMAQ Evaluation and Assessment* responds to that request by exploring different practices and approaches that select agencies Nationwide use in CMAQ project selection and implementation. The study team conducted 1-day site interviews with Metropolitan Planning Organizations (MPOs) and State Departments of Transportation (DOTs) at seven locations around the country. The *Phase II Report* highlights effective CMAQ implementation practices and identifies benefits, challenges, and opportunities encountered by the agencies interviewed as they program CMAQ funds from the information collected and analyzed during the site visit interviews.			
14. SUBJECT TERMS Congestion Mitigation and Air Quality Improvement Program (CMAQ), air quality, evaluation, assessment, cost-effectiveness, SAFETEA-LU 1808, Metropolitan Planning Organization (MPO)		15. NUMBER OF PAGES 76	
		16. PRICE CODE	
17. SECURITY CLASSIFICATION OF REPORT Unclassified	18. SECURITY CLASSIFICATION OF THIS PAGE Unclassified	19. SECURITY CLASSIFICATION OF ABSTRACT Unclassified	20. LIMITATION OF ABSTRACT Unlimited

SI* (MODERN METRIC) CONVERSION FACTORS

APPROXIMATE CONVERSIONS TO SI UNITS

Symbol	When You Know	Multiply By	To Find	Symbol
LENGTH				
in	inches	25.4	millimeters	Mm
ft	feet	0.305	meters	M
yd	yards	0.914	meters	M
mi	miles	1.61	kilometers	Km
AREA				
in^2	square inches	645.2	square millimeters	mm^2
ft^2	square feet	0.093	square meters	m^2
yd^2	square yard	0.836	square meters	m^2
ac	acres	0.405	hectares	Ha
mi^2	square miles	2.59	square kilometers	km^2
VOLUME				
fl oz	fluid ounces	29.57	milliliters	ML
gal	gallons	3.785	liters	L
ft^3	cubic feet	0.028	cubic meters	m^3
yd^3	cubic yards	0.765	cubic meters	m^3

NOTE: volumes greater than 1000 L shall be shown in m^3

MASS				
oz	ounces	28.35	grams	G
lb	pounds	0.454	kilograms	Kg
T	short tons (2000 lb)	0.907	megagrams (or "metric ton")	Mg (or "t")
TEMPERATURE (exact degrees)				
°F	Fahrenheit	5 (F-32)/9 or (F-32)/1.8	Celsius	°C
ILLUMINATION				
fc	foot-candles	10.76	lux	Lx
fl	foot-Lamberts	3.426	candela/m^2	cd/m^2
FORCE and PRESSURE or STRESS				
lbf	poundforce	4.45	newtons	N
lbf/in^2	poundforce per square inch	6.89	kilopascals	Kpa

APPROXIMATE CONVERSIONS FROM SI UNITS

Symbol	When You Know	Multiply By	To Find	Symbol
LENGTH				
mm	millimeters	0.039	inches	In
m	meters	3.28	feet	Ft
m	meters	1.09	yards	Yd
km	kilometers	0.621	miles	Mi
AREA				
mm^2	square millimeters	0.0016	square inches	in^2
m^2	square meters	10.764	square feet	ft^2
m^2	square meters	1.195	square yards	yd^2
ha	hectares	2.47	acres	Ac
km^2	square kilometers	0.386	square miles	mi^2
VOLUME				
mL	milliliters	0.034	fluid ounces	fl oz
L	liters	0.264	gallons	Gal
m^3	cubic meters	35.314	cubic feet	ft^3
m^3	cubic meters	1.307	cubic yards	yd^3
MASS				
g	grams	0.035	ounces	Oz
kg	kilograms	2.202	pounds	Lb
Mg (or "t")	megagrams (or "metric ton")	1.103	short tons (2000 lb)	T
TEMPERATURE (exact degrees)				
°C	Celsius	1.8C+32	Fahrenheit	°F
ILLUMINATION				
lx	lux	0.0929	foot-candles	Fc
cd/m^2	candela/m^2	0.2919	foot-Lamberts	Fl
FORCE and PRESSURE or STRESS				
N	newtons	0.225	poundforce	Lbf
kPa	kilopascals	0.145	poundforce per square inch	lbf/in^2

*SI is the symbol for the International System of Units. Appropriate rounding should be made to comply with Section 4 of ASTM E380. (Revised March 2003)

Table of Contents

Executive Summary v

1. **Introduction** 1
   Context for the Report: CMAQ History 1
   Purpose of the Study 1
   Study Objectives 2
   Study Methodology 2
   Report Organization 3

2. **Effective Field Practices in CMAQ Implementation** 5
   Transparent Project Solicitation, Prioritization, and Selection Processes 5
   Effective Practices from the Field 6
   Standardized Approaches to Project Evaluation and Ranking 8
   Effective Practices from the Field 11
   Adaptability in Response to Evaluations and Changing Conditions 12
   Effective Practices from the Field 13

3. **Observations from the Field: Benefits, Challenges, and Opportunities** 15
   Links to the Bigger Picture 15
   State DOT Involvement 16
   Air Quality Agency Involvement 18
   Data and Analysis Requirements 19

4. **Putting it all Together: Legislation and Guidance Translate into Projects** 21
   Birmingham Truck Stop-Electrification 22
   Statewide School Bus Retrofits in Massachusetts 24
   Denver's Traffic Signal System Improvement Program (TSSIP) 26
   Fort Collins Community Bicycle Library 28
   Medford, Oregon Diesel Retrofits 30
   Pittsburgh Downtown Car-Sharing Program 31
   Bay Area Freeway Service Patrol (FSP) 33

5. **Conclusions** 35

6. **Site-Visit Case Studies** 39
   Birmingham, Alabama: Regional Planning Commission of Greater Birmingham (RPCGB) 39
   Boston, Massachusetts: Boston MPO 42
   Denver, Colorado: Denver Regional Council of Governments (DRCOG) 44
   Fort Collins, Colorado: North Front Range MPO (NFRMPO) 47
   Medford, Oregon: Rogue Valley Council of Governments (RVCOG) 49
   Pittsburgh, Pennsylvania: Southwestern Pennsylvania Commission (SPC) 52
   San Francisco, California: Metropolitan Transportation Commission (MTC) 55

Appendix A: CMAQ Phase II Site-Visit Locations and MPO Points of Contact 59

Appendix B: CMAQ Phase II Site Visits—Interview Guide 61

Appendix C: Acronyms and Abbreviations 65

List of Tables

Table 1: Summary of Quantitative Measures to Evaluate CMAQ Proposals..9

EXECUTIVE SUMMARY

CMAQ: Strengthening National Efforts to Meet Air Quality Standards

In 1990, Congress amended the Clean Air Act (CAA) to strengthen National efforts to attain the National air quality standards. Among other strengthening provisions, the amendments required stronger coordination and linkages between transportation and air quality planning. Shortly thereafter, in 1991, Congress passed the Intermodal Surface Transportation Efficiency Act (ISTEA), which realigned the focus of transportation planning toward a more inclusive, environmentally sensitive, and multimodal approach. This included the Congestion Mitigation and Air Quality (CMAQ) Improvement Program, which was created to help fund transportation projects that reduce emissions.

CMAQ was reauthorized in the Transportation Equity Act for the 21st Century (TEA-21) in 1998, and again in 2005 with the Safe, Accountable, Flexible, Efficient Transportation Act: A Legacy for Users (SAFETEA-LU).

In 2007, in consultation with the Federal Transit Administration (FTA) and the Environmental Protection Agency (EPA), the Federal Highway Administration (FHWA) began a two-phased study as required by SAFETEA-LU Section 1808(f). Section 1808(f) calls for an evaluation and assessment of the direct and indirect impacts of CMAQ-funded projects on air quality and congestion levels to ensure the program's effective implementation.

The following report presents results from the second phase of that study and should be read in conjunction with *SAFETEA-LU 1808: CMAQ Evaluation and Assessment, Phase I Final Report*.[1] The primary objectives of the Phase II study were to:

- Explore practices and approaches to CMAQ project selection and implementation.
- Provide information for use by local, regional, and State transportation agencies for the purpose of ensuring effective CMAQ implementation.

Using Case Studies to Explore CMAQ Project Selection and Implementation

As part of the Phase II research, a Federal CMAQ Team conducted 1-day interviews with Metropolitan Planning Organizations (MPOs) and State Departments of Transportation (DOTs) at the following seven locations Nationwide:

- Birmingham, Alabama
- Boston, Massachusetts
- Denver, Colorado
- Fort Collins, Colorado
- Medford, Oregon
- Pittsburgh, Pennsylvania
- San Francisco, California

This report highlights effective practices and identifies benefits, challenges, and opportunities encountered by agencies as they program CMAQ funds from the information collected and analyzed at these site visits. The exploration of effective practices may be useful and instructive to other MPOs and State DOTs around the country as they build their own experience with the CMAQ program and seek to improve their project implementation.

---

[1] *SAFETEA-LU 1808: Congestion Mitigation and Air Quality Improvement Program Evaluation and Assessment, Phase I Final Report;* available at: http://www.fhwa.dot.gov/environment/cmaqpgs/safetealu1808/safetealu1808.pdf.

Effective Practices—Highlights

Processes for selecting, programming, and evaluating CMAQ projects varied by site location so effectiveness was broadly defined to reflect the range of circumstances observed in the field. Effectiveness was also assessed in relation to FHWA's CMAQ program guidance and results from Phase I of the study. Effective practices were identified in the three areas highlighted below.

Transparent project solicitation, prioritization, and selection processes

Two broad types of transportation planning processes were used for evaluating and selecting CMAQ projects, and good practices were found within each. The two process types were:

- A stand-alone CMAQ process, with its own CMAQ-specific outreach efforts.
- An integrated process, where projects are solicited, evaluated, and selected as part of an overall planning and programming process.

Standardized approaches to project evaluation and ranking

MPOs developed a variety of methodologies for project evaluation. Effective practices were found in the following areas:

- Developing quantitative and qualitative measures.
- Estimating air quality or congestion benefits.
- Determining cost-effectiveness.
- Evaluating and ranking projects.

Adaptability in response to evaluations and changing conditions

One success factor to developing a good CMAQ program was found to be an organization's willingness and ability to adapt and change in response to a variety of shifting factors. Those factors included evolving local conditions, new Federal guidance/legislation, and the results of project outcome evaluations. Effective practices were found in the following areas:

- Conducting a periodic, broad-level assessment of the CMAQ program.
- Conducting post-project analysis and evaluation of CMAQ projects.

Benefits, Challenges, and Opportunities

The site visits also provided a better understanding of the broader planning and funding context within which transportation agencies develop CMAQ programs. MPOs reported several benefits, challenges, and opportunities related to their CMAQ program development. For example, MPOs reported that the CMAQ program improved their capabilities to:

- Increase their planning capacity overall due to the oversight and operations required at the MPO level.
- Test innovative projects on a small scale for demonstrable success before scaling up.
- Allow broader policies to evolve that reflect the nexus between transportation and other issues.
- Open the transportation planning and programming process to nontraditional partners.
- Build stronger relationships with and obtain involvement from air quality agencies.

The CMAQ program's requirements to estimate and document project benefits presented common challenges. For example, some MPOs noted that data requirements associated with CMAQ added to project costs. On the other hand, these requirements had a positive side in that they brought a higher degree of analytical rigor and justification to the planning and programming process. Other MPOs suggested that just meeting the minimum analytical requirements did not provide accurate estimates of emissions reductions over the life of the project. Still others noted that it was a challenge to allocate the staff time required for administering CMAQ to a funding source that represents a small portion of their overall transportation budgets.

Many agencies agreed that CMAQ's additional requirements can be seen as positive at a broad level and as a model for using analysis to identify projects with clearly defined and measurable benefits.

Conclusions

While the CMAQ program is Federally funded, no National standard or set of regulations exists for how a CMAQ program should be structured and operated at the State or MPO level. It is intentionally left to the State or MPO to develop a process and operate a program that best responds to local or regional needs. Reflecting this, the Federal CMAQ Team found differences in structure and operations at each of the seven sites visited.

The agencies interviewed for this study represent a very small sample of CMAQ programs across the country but allowed the Federal CMAQ Team to draw several conclusions about potential effective practices as well as benefits, challenges, and opportunities presented by the CMAQ program. These are:

- Transparent project solicitation, prioritization, and selection processes play important roles in engaging citizens and increasing stakeholder involvement.
- Standardized processes to evaluate and rank multiple project types are important for increasing transparency and giving public confidence, but can be challenging especially due to CMAQ's emphasis on nontraditional projects.
- Effective processes for evaluating program impact and the ability to adapt in response to evaluations and/or changing conditions are important for continual program improvement.
- State DOTs play an important role in shaping CMAQ program development at the MPO level.
- CMAQ engages air quality agencies as valuable partners in the transportation planning process.
- CMAQ's data and analysis requirements present both significant challenges and opportunities.
- CMAQ legislation and guidance are translated and applied to meet local transportation and air quality requirements and concerns, resulting in a range of innovative projects.
- Positive impacts go beyond stated CMAQ program goals.

To demonstrate by specific examples the range and variety of CMAQ projects that have arisen from the legislation and guidance, this report includes seven brief summaries of CMAQ-funded projects. The summaries highlight projects that the MPOs interviewed felt were successful in areas such as providing air quality benefits, strengthening interagency cooperation, increasing economic vitality, improving quality of life, and leveraging funds to maximize impact across geographic boundaries. The summaries will help readers to gain insight into the benefits, challenges, and opportunities reported by agencies involved with CMAQ project selection and evaluation.

1. INTRODUCTION

Context for the Report: CMAQ History

A short history of the Congestion Mitigation Air Quality (CMAQ) Improvement Program provides an important context for this report. In 1990, Congress amended the Clean Air Act (CAA)[2] to strengthen National efforts to attain the National Ambient Air Quality Standards (NAAQS). The amendments required reductions in tailpipe emissions, initiated stronger control measures in areas that failed to attain the NAAQS (nonattainment areas [NAAs]), and provided for a stronger connection between transportation and air quality planning. Shortly thereafter, in 1991, the CMAQ program was created with the Intermodal Surface Transportation Efficiency Act[3] (ISTEA) in order to realign the focus of transportation planning toward a more inclusive, environmentally sensitive, and multimodal approach to addressing transportation problems.

The CMAQ program was reauthorized in the Transportation Equity Act for the 21st Century (TEA-21)[4] in 1998 and again in 2005 with the Safe, Accountable Flexible, Efficient Transportation Equity Act: A Legacy for Users (SAFETEA-LU). Each of these subsequent bills resulted in a shift of priorities within the CMAQ program at the Federal, State, and regional levels. A provision within SAFETEA-LU established priority consideration for cost-effective emissions reduction and congestion mitigation activities. Between Fiscal Years (FYs) 2005 and 2009, CMAQ represented 4.3 percent of the Federal Highway Administration's (FHWA) total authorizations Nationwide.[5] In October 2008, FHWA issued Final Program Guidance (http://www.fhwa.dot.gov/environment/cmaq08gm.htm) for the CMAQ program.[6] The guidance notes that the CMAQ program supports two important goals of the U.S. Department of Transportation (USDOT): improving air quality and relieving congestion. In addition, the guidance provides information on project eligibility, project selection, and program administration.

Purpose of the Study

This report documents the second part of a two-phased study of the CMAQ program. The study was led by FHWA in consultation with the Federal Transit Administration (FTA) and the Environmental Protection Agency (EPA). It was undertaken in response to requirements in Section 1808(f) of SAFETEA-LU (Pub. L. 109-59, Aug. 10, 2005) to evaluate and assess the CMAQ program.[7] The text of this legislation is as follows:

> (f) EVALUATION AND ASSESSMENT OF CMAQ PROJECTS.—Section 149 of such title (as amended by subsection (e)) is amended by adding at the end the following:
> (h) EVALUATION AND ASSESSMENT OF PROJECTS.—
> (1) IN GENERAL.—The Secretary, in consultation with the Administrator of the Environmental Protection Agency, shall evaluate and assess a representative sample of projects funded under the congestion mitigation and air quality program to—
> (A) determine the direct and indirect impact of the projects on air quality and congestion levels; and

---

[2] *Clean Air Act 1990 Amendment;* available at http://www.epa.gov/air/caa.
[3] *Intermodal Surface Transportation Efficiency Act (ISTEA) of 199,* Public Law 102-240; available at http://www.bts.gov/laws_and_regulations.
[4] *Transportation Equity Act for the 21st Century (TEA-21) of 1998,* Public Law 105-178; available at http://www.bts.gov/laws_and_regulations.
[5] *Safe, Accountable, Flexible, Efficient Transportation Equity Act: A Legacy for Users (SAFETEA-LU),* Public Law 109-59, Funding Tables, August 2005; available at http://www.fhwa.dot.gov/safetealu/fundtables.htm.
[6] Federal Highway Administration. *The Congestion Mitigation and Air Quality (CMAQ) Improvement Program under the Safe, Accountable, Flexible, Efficient Transportation Equity Act: A Legacy for Users, Final Program Guidance;* available at http://www.fhwa.dot.gov/environment/cmaq08gm.htm.
[7] *Safe, Accountable, Flexible, Efficient Transportation Equity Act: A Legacy for Users (SAFETEA-LU),* Public Law 109-59, August 2005; available at http://frwebgate.access.gpo.gov/cgi-bin/getdoc.cgi?dbname=109_cong_public_laws&docid=f:publ059.109.

(B) ensure the effective implementation of the program.
(2) DATABASE.—Using appropriate assessments of projects funded under the congestion mitigation and air quality program and results from other research, the Secretary shall maintain and disseminate a cumulative database describing the impacts of the projects.
(3) CONSIDERATION.—The Secretary, in consultation with the Administrator, shall consider the recommendations and findings of the report submitted to Congress under section 1110(e) of the TEA-21 (112 Stat. 144), including recommendations and findings that would improve the operation and evaluation of the congestion mitigation and air quality improvement program.

This report should be read in conjunction with SAFETEA-LU 1808: *CMAQ Evaluation and Assessment: Phase I Final Report.*[8]

Study Objectives

Phase II of the CMAQ Evaluation and Assessment Study was designed to respond to Congress's direction to ensure the effective implementation of the program. It was also designed to provide important information about CMAQ implementation to Metropolitan Planning Organizations (MPOs) and State Departments of Transportation (DOTs), which can use the information highlighted to assess and improve their programs.

Study Methodology

Phase I of the study evaluated 67 CMAQ-funded projects from the FHWA database, using emissions and cost-effectiveness data in response to the SAFETEA-LU requirement to determine the impacts of a sample of CMAQ projects. The Phase I report noted a variety of good practices that States and MPOs used to analyze and program CMAQ projects. These practices included:

- Development of standardized methods to calculate benefits.
- Development of procedures for ranking projects, including consideration of cost-effectiveness.
- Coordination with air and local agencies in the project selection process.

Phase II of the study responds to the SAFETEA-LU requirement to ensure effective implementation of the CMAQ program. The FHWA/EPA study team, referred to as the Federal CMAQ Team throughout this report, conducted Phase II by selecting MPOs that were implementing some or all of these practices and then conducting interviews and developing case studies to document effective CMAQ project selection and implementation practices in the field. The Phase II report also provides observations from the field about challenges and opportunities for improvement in CMAQ program implementation.

The Federal CMAQ Team conducted 1-day site visits to seven locations across the country between June and September 2008. The seven locations were chosen on the basis of criteria and findings from the Phase I study. The Federal CMAQ Team identified sites that (1) had relatively high obligation rates for CMAQ funds and (2) warranted deeper analysis because they were found to have one or more of the following characteristics identified as good practices during the Phase I analysis:

- Transparent project prioritization/selection processes.
- Standardized tools or emissions calculation methods.
- Effective processes for evaluating project or program impact.

In choosing the locations for site visits, the Federal CMAQ Team also considered both geographic and MPO size diversity and included some locations that had experience with diesel engine retrofit projects,

---

[8] *SAFETEA-LU 1808: CMAQ: Phase I Final Report;* available at http://www.fhwa.dot.gov/environment/cmaqpgs/safetealu1808/safetealu1808.pdf.

since this type of project was specifically encouraged under the SAFETEA-LU legislation due to its air quality benefits. The seven sites were:

- Birmingham, Alabama
- Boston, Massachusetts
- Denver, Colorado
- Fort Collins, Colorado
- Medford, Oregon
- Pittsburgh, Pennsylvania
- San Francisco, California

Appendix A provides more detail on the points of contact for each site visit agency, along with the locations of their regional offices of FTA, FHWA, and EPA.

The Federal CMAQ Team developed and used a standard questionnaire for all site visits. The questions and agendas were sent to the State, regional, and local representatives of each MPO prior to the visit. Site-visit participants included staff from the MPOs and State DOTs as well as representatives from one or two local CMAQ project-sponsoring agencies. Where possible, participants also included representatives from the State or regional air quality/natural resource agencies. Federal staff in attendance included representatives from FHWA Headquarters and field offices, EPA Headquarters and regional offices, FTA regional offices, and USDOT's Volpe National Transportation Systems Center (Volpe Center). The questions, which can be found in Appendix B, focused on the following areas:

- CMAQ program objectives.
- CMAQ program procedures, including project initiation, analysis, and selection.
- CMAQ program and project funding.
- Reporting.
- Program and project evaluation.
- In-depth description of at least two recently programmed CMAQ projects.

The seven case studies are not intended to provide a statistically significant representation of the overall CMAQ program Nationally. Rather, they provide examples of the challenges facing agencies as they program CMAQ funds as well as examples of how agencies have effectively responded to Federal legislation and guidance as they develop methods to program and evaluate projects to meet the program's goals. The MPOs served as the main points of contact for the site visits. The term MPO is used throughout the document as opposed to region, agency, or another similar descriptor.

Report Organization

The report is organized into six major sections and three appendices:

- Section 1: Introduction
- Section 2: Effective Field Practices in CMAQ Implementation
- Section 3: Observations from the Field: Benefits, Challenges, and Opportunities
- Section 4: Putting It All Together: Legislation and Guidance Translate into Projects
- Section 5: Conclusions
- Section 6: Site-Visit Case Studies
- Appendix A: CMAQ Phase II Site-Visit Locations and Points of Contact
- Appendix B: CMAQ Phase II Site Visits—Interview Guide
- Appendix C: Acronyms and Abbreviations

2. EFFECTIVE FIELD PRACTICES IN CMAQ IMPLEMENTATION

There is no single National standard or set of regulations dictating how a CMAQ program should be structured and operated at the State or MPO level, and the CMAQ authorizing legislation does not require the issuance of such regulations. FHWA offers CMAQ program guidance, but intentionally leaves discretion to MPOs and State DOTs to develop a process that best responds to needs at the regional or local level. Accordingly, the Federal CMAQ Team found differences in structure and operations at every one of the sites visited. Processes for selecting, programming, and evaluating CMAQ projects are highly diverse from region to region and may even vary within a single State. Agencies also varied in their strengths as well as in the challenges they faced to effective CMAQ implementation.

This section of the report documents effective CMAQ implementation practices found in the seven field locations where interviews were conducted. The Federal CMAQ Team identified effective practices across a diverse set of agencies, where each agency is working to implement a CMAQ program that best responds to local needs and conditions. Effectiveness was assessed in relation to FHWA's CMAQ program guidance and the practices identified during the Phase I analysis. Generally, effective CMAQ programs are those that exhibit some or all of the following characteristics:

- Transparent project solicitation, prioritization, and selection processes.
- Standardized approaches to project evaluation and ranking.
- Adaptability in response to evaluations and changing conditions.

The exploration of effective practices and noteworthy examples described below may be useful and instructive to other MPOs around the country as they build their own experience with the CMAQ program and seek to improve their own implementation of CMAQ funded projects. Readers are encouraged to delve more deeply by reading the case study of each site visit (included in Section 6) to identify which practices may be applicable to their own organizations and to contact the organizations for information-sharing.

Transparent Project Solicitation, Prioritization, and Selection Processes

Federal guidance for the CMAQ program states the need for a selection process that is "transparent, in writing, and publicly available."[9] How MPOs or States structure their CMAQ process may vary widely, but in every case it is important that there be clearly understood, documentation and guidance on:

- The operation of the CMAQ program.
- How projects are solicited and proposed.
- Who is able to propose a project.
- How projects are evaluated and ranked.
- How funding decisions are made.

In line with Federal guidance and the findings from Phase I, information gathered during the Phase II study confirmed that adopting transparent project solicitation, prioritization, and selection processes is a critical success factor for effectively implementing a good CMAQ program. Common attributes of open, understandable CMAQ program processes were:

- The program's vision, goals, solicitation, evaluation, and selection processes are clear, understandable, reliable, and documented.

---

[9] Federal Highway Administration. *The Congestion Mitigation and Air Quality (CMAQ) Improvement Program under the Safe, Accountable, Flexible, Efficient Transportation Equity Act: A Legacy for Users, Final Program Guidance*, p. 29; available at http://www.fhwa.dot.gov/environment/cmaq08gm.htm.

- The program's vision, goals, solicitation, evaluation, and selection processes are easily accessible by the public and supported by strong education and outreach efforts.
- The operation of the program and the decisionmaking process are transparent and reliable.

The above characteristics are important to gaining and maintaining trust by policy-makers and the public. Greater understanding of, familiarity with, and trust in MPO processes by potential project sponsors (i.e., applicants to the CMAQ program) translates into more project submittals, better competition, and ideally, the selection of better projects for funding. When project sponsors do not understand how projects rise to the top of the selection process, they may feel that decisions are made for nonmerit reasons, thereby undermining community trust in the MPO planning process.

The results from the site interviews suggest that the public process used to solicit and program CMAQ projects is closely tied to an MPO's overall approach to programming Federal transportation funds. Approaches typically fall in one of two major categories:

1. A stand-alone CMAQ process with its own CMAQ-specific outreach efforts, as found in:
   - Birmingham, Alabama
   - Fort Collins, Colorado
   - Medford, Oregon
   - Pittsburgh, Pennsylvania

2. An integrated process where CMAQ monies are pooled with other funding sources and all projects are solicited, evaluated, and selected as part of an overall planning and programming process, as found in:
   - Boston, Massachusetts
   - Denver, Colorado
   - San Francisco, California

Both approaches can support effective CMAQ implementation. Of the seven agencies interviewed for this report, smaller MPOs, with smaller budgets, fewer staff, and relatively less severe air quality challenges, tended to run CMAQ as a stand-alone program. The larger MPOs, with longer-standing programs and more complex congestion mitigation and air quality improvement challenges, tended to integrate CMAQ into the overall planning process. However, whether an MPO runs CMAQ as a stand-alone program or treats it as just one funding source among many, every State and MPO must integrate projects programmed through CMAQ into its regional Transportation Improvement Program (TIP) and then its State Transportation Improvement Program (STIP). As such, CMAQ project solicitation typically follows the same schedule as TIP updates.

Effective Practices from the Field

CMAQ Program Guidance and Consolidated CMAQ Calendar

North Front Range Metropolitan Planning Organization (NFRMPO), Fort Collins, Colorado

In Fort Collins, NFRMPO has produced and put on its Web site a 41-page document that describes the purpose of the CMAQ program, enumerates the agency's priorities for using CMAQ funds, lists project eligibility, and details the project selection process.[10] This clear outline of the CMAQ program goals, operating procedures, and selection process allows an interested citizen or organization with a project idea to read and understand how to get involved. In addition, the Web site has a consolidated calendar for CMAQ projects under consideration for the FY 2010–2011 funding cycle.

---

[10] North Front Range Metropolitan Organization (NFRMPO). *Project Submittal and Evaluation Process Guidebook for FY 10–11*, March 2009; available at: http://www.nfrmpo.org/Projects/CCFP.aspx.

Reformulation of the CMAQ Process

Southwestern Pennsylvania Commission (SPC), Pittsburgh, Pennsylvania

In order to improve the openness and transparency of the CMAQ selection process for project sponsors, planning partners, and the public, the SPC board and staff worked to reformulate the Pittsburgh region's CMAQ process. In 2007, SPC hired an independent facilitation team to assist with the CMAQ review to help ensure a fair and unbiased process. As a result, SPC updated its CMAQ evaluation and selection procedures and established a new CMAQ Evaluation Committee (CEC) to help review applications, prioritize projects, and make recommendations on which projects to fund. The CEC is composed of more than 20 members representing SPC's member counties, transportation management associations (TMAs), the Pennsylvania Department of Transportation (PennDOT), local environmental agencies, and SPC modal committees (freight forum, pedestrian/bicycle, transit operators). The revised process was endorsed by the SPC board and used to develop the recommended CMAQ Program for the 2009–2012 TIP.

Online Application Process

Southwestern Pennsylvania Commission (SPC), Pittsburgh, Pennsylvania

SPC developed a revised CMAQ application that project sponsors can download from the Web, complete, and return to SPC electronically. The application is an interactive PDF file that allows project sponsors to attach documents or maps. It includes detailed instructions for applicants and provides a description of the project evaluation criteria used in the scoring, ranking, and selection process. SPC spent considerable effort in updating the application questions to ensure that all pertinent information is supplied for each project and to provide background information about the CMAQ program and guidance to help applicants understand the criteria that will be used to evaluate their projects.

Diesel Retrofit Public Outreach

Rogue Valley Metropolitan Planning Organization (RVMPO), Medford, Oregon

When the RVMPO Policy Committee decided that diesel retrofit projects should be a priority for how to spend CMAQ funding, RVCOG staff organized a public open house specifically to reach out to local businesses operating diesel fleets in the area. The open house received media coverage from two television stations and a newspaper. The open house informed companies that Federal funding was available to support retrofitting their fleets' engines to burn cleaner, provided educational materials to show that retrofits do not adversely affect vehicle operations, and gave tips on how to apply for a CMAQ grant. A product of this outreach led to a successful public-private partnership (PPP) with a private company in the area, Rogue Disposal & Recycling, to begin retrofitting its fleet.

Air Quality Outreach

Regional Planning Commission of Greater Birmingham (RPCGB), Birmingham, Alabama

The CMAQ program's emphasis on innovative and nontraditional projects provides an opportunity for MPOs to foster partnerships with organizations that have not typically been involved in the metropolitan transportation planning process, such as air quality agencies, community organizations, and private firms. In Birmingham, RPCGB dedicates about $1.4 million of CMAQ funding annually to support the work of the Alabama Partners for Clean Air (APCA), a consortium of 14 public, private, and nonprofit organizations working to implement projects and programs that improve air quality in the RPCGB area. As such, the CMAQ program helps RPCGB to broaden public awareness of the metropolitan planning process by reaching out to and partnering with a wider range of organizations than it might have otherwise.

Standardized Approaches to Project Evaluation and Ranking

One of the distinguishing features of the CMAQ program is that project sponsors must estimate their proposed project's air quality benefits in order to be eligible for Federal funding. As described in the CMAQ Program Final Guidance issued in November 2008, agencies are expected to give priority consideration to cost-effective projects.[11] This requires that MPOs and State DOTs have the capacity to develop multiple methodologies to quantify outcomes and then apply cost-benefit analyses across a wide range of project categories and types. One of the most effective ways that agencies can accomplish these objectives is to standardize approaches. Developing standardized calculations also helps to clarify the methodologies by which CMAQ project proposals are ranked and selected, thereby increasing the transparency and openness of the programming process.

Site-visit interviews revealed that, while each agency had developed some type of standardized calculations, no two methodologies were the same. Several common components were identified, however, that helped to make the development and use of standardized methodologies more effective. These components were:

1. Developing quantitative and qualitative measures.
2. Estimating air quality or congestion benefits.
3. Determining cost-effectiveness.
4. Evaluating and ranking projects.

Developing quantitative and qualitative measures: Agencies use both quantitative and qualitative measures to estimate project costs and benefits and to evaluate CMAQ proposals. The development and choice of which measures to use is typically an iterative process, with the MPO board providing policy guidance and the staff developing the analytical formulas on which quantitative measures are based. Each MPO interviewed had its own unique set of quantitative and qualitative measures that it used to evaluate project submissions.

Quantitative measures include assumptions and standardized formulas that are tailored to regional characteristics, needs, and goals and used to calculate project benefits. Common quantitative measures among agencies interviewed were:

- Emissions reduction.
- Reduction in number of automobile trips.
- Change in vehicle-miles traveled (VMT).
- Cost-effectiveness (e.g., per-unit reduced emissions, per-unit change in reduced number of automobile trips, per-unit change in VMT).

In some cases, State DOTs provide technical support to MPOs to develop standardized calculation methodologies. For example, while CMAQ projects are programmed at the MPO level in Pennsylvania, PennDOT has a statewide consultant who creates spreadsheets that can be used to calculate estimated air quality benefits for a wide range of CMAQ-type projects. The consultant remains on call to help MPOs with new or nonstandard project proposals (i.e., those for which a methodology has not yet been created). In other areas, the MPO relies on its own staff or an outside consultant to help formulate the quantitative measures. Table 1 provides a summary of how each of the seven MPOs interviewed developed its existing quantitative calculation measures. (More detailed information is included in Section 6: Site Visit Case Studies.)

---

[11] Federal Highway Administration. *The Congestion Mitigation and Air Quality (CMAQ) Improvement Program under the Safe, Accountable, Flexible, Efficient Transportation Equity Act: A Legacy for Users, Final Program Guidance;* available at http://www.fhwa.dot.gov/environment/cmaq08gm.htm.

Table 1: Summary of Quantitative Measures to Evaluate CMAQ Proposals

Boston MPO	Standardized spreadsheet developed by State DOT
Birmingham (RPCGB)	Standardized formulas developed by consultant
Denver (DRCOG)	Standardized formulas developed by MPO
Fort Collins (NFRMPO)	Standardized formulas developed and refined by consultant
Medford (RVCOG)	Standardized calculations performed by MPO
Pittsburgh (SPC)	Standardized formulas developed by Statewide consultant
San Francisco (MTC)	Supplements State calculations with its own MPO work

Qualitative measures are more subjective and difficult to standardize. They are typically criteria that respond to MPO policy board goals by awarding points to proposals during the project evaluation process to encourage:

- Desired project types (e.g., diesel retrofits, new/innovative projects).
- Projects in particular geographic areas (e.g., within urban growth boundary or transit corridor).
- Multimodal system connectivity (e.g., intermodal centers).
- New project partners (e.g., air quality agencies, private sector, or community-based organization).

Qualitative measures can be built into the evaluation and ranking process by giving bonus points either upfront in the analysis of benefits or toward the end. They usually are developed on the basis of policy goals and popular support for certain types of projects or programs that support the region's long-range transportation vision but whose quantitative benefits are harder to estimate. For example, a number of MPOs noted that they use CMAQ funds for bicycle and pedestrian projects because they enjoy broad public and political support within the MPO even though they may not compete as successfully as other project types on a strictly quantitative basis, such as those focused on traffic signalization or intersection improvements.

Estimating air quality or congestion benefits: Once an MPO agrees on which quantitative and qualitative measures and methods to use, project benefits can be quantified. Most MPOs require that project sponsors provide the assumptions and related information needed to calculate a project's projected benefits; staff then provide a review to verify the assumptions and results. In Pittsburgh, for example, a project sponsor is able to download an electronic application, use the online tools to calculate benefits, and then submit the completed application to the MPO for review. Other MPOs perform all calculations in-house, using information provided by project sponsors on their applications.

In interviews, agencies emphasized that the use of quantitative analysis to summarize the benefits of projects helps to validate the selection process with the public and project sponsors by allowing everyone to see the calculated benefits. However, several agencies cautioned that the quality of benefits calculations is only as good as the assumptions built into the formulas. While a formula for a certain type of project may be applicable at the State or National level, many of the default variables need to be changed to reflect conditions that exist at the local or regional level.

In addition to the question of how to calculate benefits is the question of which benefits to calculate from the standpoint of improving air quality regionwide and what timeframe to use. Most MPOs follow FHWA guidance that only requires the calculation of benefits in Year 1 of project implementation. The timeframe that is used impacts the types of projects that rise to the top in evaluation. For example, an intersection improvement may show a relatively high benefit in Year 1, but this benefit may erode over time. A bike lane, new bus service, or outreach and educational program, on the other hand, may not show any quantifiable benefits in Year 1 but may have benefits that accrue and build over time. The timeframe of analysis will impact whether a project appears good or not, while the underlying reality may be more subtle or context-dependent. In the State of Colorado, the DOT and MPOs are addressing this challenge by calculating life-cycle benefits for projects for inclusion in the State CMAQ Reporter in addition to the Year 1 calculations reported to FHWA for inclusion in the CMAQ database.

Determining cost-effectiveness: SAFETEA-LU directs States and MPOs to give priority to (1) diesel retrofits and other cost-effective emissions reduction activities and (2) cost-effective congestion mitigation activities that provide air quality benefits. The determination of cost-effectiveness requires calculation of both the benefits and costs for each project. This is usually expressed as the amount in kilograms of air pollutant reduced divided by the cost of the project. In Medford, Oregon, project cost-effectiveness accounts for 16 percent of the total possible score during the project evaluation phase. The next section, on effective practices, provides greater detail on how the RVCOG in Medford awards points in determining which projects to fund.

A separate cost-effectiveness measurement is usually calculated for each of the identified pollutants within the air quality district. Different categories of projects are typically better at reducing different types of pollutants. For example, the *CMAQ Evaluation and Assessment Phase I Final Report* found that diesel retrofits can be some of the most cost-effective projects in reducing both ozone precursors and particulate matter (PM).[12] Traffic flow improvements can be cost-effective in reducing carbon monoxide (CO) and ozone precursors. In the Pittsburgh area, SPC utilizes a set of standardized models to compare benefits and costs in order to develop a cost-effectiveness rating for each CMAQ candidate.

Evaluation and ranking of projects: Once the quantitative and qualitative measures for each project are documented, the evaluation and ranking of projects can be accomplished. Most MPOs incorporate qualitative measures into this step by giving bonus points for certain types of projects. The stated and publicized decision to favor certain types of projects over others gives project sponsors important information about what types of projects are likely to be funded. Each of the seven locations studied develops its own evaluation methodology, with different combinations of points and mixes of qualitative and quantitative measures. This flexibility enables tailoring of the program to meet local needs.

Once projects have been evaluated and ranked, the MPO policy board or its designated committee will typically review the list of eligible CMAQ projects to ensure a balanced funding program. This review will often look at the geographic distribution of project proposals and also will check to ensure that regional policy goals are met by the mix of recommended proposals. MPOs with many project submissions will often organize applications within project-type categories and then reevaluate and rank proposals within these categories to ensure fair and logical comparisons. These agencies have developed multiple evaluation methodologies tailored to each project-type category since, for example, a bicycle project has different assumptions and characteristics than an intersection improvement or diesel retrofit project.

Several of the MPOs have revised the types of project categories that are emphasized over the years to better serve changing needs. In Medford, for example, paving projects were highly ranked with use of quantitative measures because they led to quantifiable reductions in PM, which was the MPO's major air quality challenge. However, when diesel retrofits became a CMAQ funding priority in SAFETEA-LU, the board decided to emphasize these types of projects and revised its ranking process by awarding points explicitly for diesel retrofits.

---

[12] *SAFETEA-LU 1808: Congestion Mitigation and Air Quality Improvement Program Evaluation and Assessment: Phase I Final Report*, p. 50; available at: http://www.fhwa.dot.gov/environment/cmaqpgs/safetealu1808/safetealu1808.pdf.

Effective Practices from the Field

Sample Project Evaluation Framework

Rogue Valley Council of Governments (RVCOG), Medford, Oregon

Project Evaluation Criteria

Staff work with members of the Technical Advisory Committee (TAC) to review project proposals and to score and rank proposals with use of a single set of criteria. Each project can earn up to a total of 115 points on the basis of both quantitative and qualitative factors organized within 10 criteria categories that reflect the Policy Committee's CMAQ priorities.

Project Evaluation Criteria	Maximum Points	Points Awarded
Project results in CO/PM 10 reduction (contribution to overall reduction)	20	
Project is cost-effective (benefit/cost = kg reduction/ $ spent)	20	
Project results in long-term air quality improvement (effectiveness in 5 years, 10 years, etc.)	15	
Project demonstrates potential to reduce reliance on automobiles	15	
Project demonstrates potential to mitigate congestion	15	
Project helps to complete a multimodal transportation system	10	
Diesel retrofit project	5	
Project is in city limits or inside Urban Containment Boundary	5	
Project results in reduction of nitrogen oxide (NO_x) and/or volatile organic compounds (VOCs)	5	
Innovation (e.g., new to the area, new technology)	5	

Standardized Air Quality Worksheets

Massachusetts Executive Office of Transportation (EOT)

The Massachusetts EOT developed a series of air quality analysis worksheets to be used by each of the State's MPOs to evaluate CMAQ-eligible projects. The worksheets incorporate the emissions factors for each of the two air quality regions within the State and allow for changes to input variables, such as average speed and trip length, depending on regional conditions. Outputs of the worksheets include net emissions change and cost-effectiveness calculations for each type of emission. This allows for a comparison of project cost-effectiveness on a statewide basis.

Evaluation of Project Benefits

Southwestern Pennsylvania Commission (SPC), Pittsburgh, Pennsylvania

At the Pittsburgh MPO, each CEC member is asked to evaluate the assumptions, calculations, and qualitative benefits submitted with applications for each project. The review imposes a reality check on initial project submissions to ensure greater accuracy and a better understanding of each project by the CEC members. SPC staff also support project review by quantifying each proposed project's impact on air quality, using a standardized set of analysis tools developed by PennDOT. Evaluation criteria include:

- Change in emissions.
- Change in VMT.
- Change in trips.
- CMAQ cost-per-unit change in emissions.
- CMAQ cost-per-unit change in trips and VMT.

The CMAQ evaluation process also considers a set of ancillary factors to assess each project's deliverability, local commitment, and consistency with the policies and goals in the region's long-range plan, county and local comprehensive plans, and the region's congestion management process.

Calculation of Life-Cycle Benefits
Colorado Department of Transportation (CDOT)

In 2000, the Colorado Transportation Commission expressed concern about the effectiveness of the CMAQ program in improving air quality and adopted a resolution (TC-807) to increase accountability for how CMAQ funds are spent. This led to the development of the CMAQ Reporter, a database maintained by CDOT that requires fund recipients to report annually on the effectiveness of their CMAQ expenditures. As a result, MPOs within Colorado are required to calculate the air quality benefits that will accrue over the lifetime of a CMAQ project.

CMAQ Evaluation Committee
North Front Range MPO (NFRMPO), Fort Collins, Colorado

NFRMPO has clearly established CMAQ evaluation criteria and an easy-to-understand, three-tiered scoring system to select projects. This includes short-term air quality benefits in Year 1 (50 percent of score), long-term air quality benefits in Years 2 to 5 (20 percent of score), and regional planning achievement (30 percent of score). Because the process of demonstrating air quality benefits can be contentious between transportation and environmental agencies, NFRMPO set up a CMAQ Evaluation Committee that includes representatives of both EPA and the Colorado Department of Public Health and Environment (CDPHE) as voting members. Using NFRMPO's established CMAQ evaluation criteria, the 14-member committee scores and then ranks proposed CMAQ projects for funding.

Adaptability in Response to Evaluations and Changing Conditions

Another important success factor in developing a good CMAQ program is an organization's willingness and ability to reflect on, and change in response to, shifting local needs and conditions, new Federal guidance/legislation, or the evaluation results of project outcomes post-implementation.

Since the CMAQ program was first created by ISTEA[13] in 1991, each of the subsequent Federal transportation bills, TEA-21[14] and SAFETEA-LU[15], has resulted in a shift of priorities within the program at the Federal, State, and regional levels. In order to effectively implement CMAQ, agencies had to be able to adapt to these changes. Several of the MPOs interviewed also discussed the value of conducting a broad-level assessment of the CMAQ program overall on a periodic basis to ensure that it supports shifting regional conditions and evolving regional policy goals.

---

[13] *Intermodal Surface Transportation Efficiency Act (ISTEA) of 199,* Public Law 102-240; available at http://www.bts.gov/laws_and_regulations.
[14] *Transportation Equity Act for the 21st Century (TEA-21) of 1998*, Public Law 105-178; available at http://www.bts.gov/laws_and_regulations.
[15] *Safe, Accountable, Flexible, Efficient Transportation Equity Act: A Legacy for Users (SAFETEA-LU),* Public Law 109-59, August 2005; available at http://frwebgate.access.gpo.gov/cgi-bin/getdoc.cgi?dbname=109_cong_public_laws&docid=f:publ059.109.

Unlike other Federal transportation funding categories, projects using CMAQ funding are required to demonstrate air quality benefits in order to be eligible. Establishing CMAQ eligibility for proposed projects requires that agencies make assumptions upfront about air quality and congestion impacts as well as project costs and benefits. Given that CMAQ supports projects that are nontraditional or completely new to an MPO, however, it can be difficult for agencies to define the correct assumptions and to accurately predict impacts. To better understand actual project benefits, some of the MPOs interviewed have introduced post-project analysis and evaluation of CMAQ projects. Evaluation results help MPOs to refine assumptions and estimates for future projects and provide policymakers with better information for future decisions. Post-project evaluation allows historical benchmarks to be set and tracked for actual, rather than projected, congestion mitigation or air quality benefits and costs. These benchmarks make it easier for new sponsors to make assumptions and develop estimates for related nontraditional transportation projects. This practice also allows the MPO to show members of the public that they are receiving a demonstrable benefit from their tax dollars.

Several MPOs noted that post-project evaluation of CMAQ projects is being driven by project sponsors. In both Pittsburgh and Birmingham, it was noted that if CMAQ project sponsors want to be able to compete successfully in the next round of CMAQ funding, the sponsor must show that its previous projects achieved the projected benefits. In Boston, it is up to the sponsor of operations-related projects to undertake usage studies and report on the benefits. The project's future level of funding depends on demonstrating benefits commensurate with the usage projections. In the long run, this may favor projects for which benefits are easier to calculate upfront (e.g., diesel retrofits) over those in which benefits build over time (e.g., bicycle and pedestrian facilities, transit, and public outreach).

Post-project evaluation is expensive both in staff time and money, so it is sometimes seen as a luxury or as being done at the expense of other important work conducted at the MPO. While it is clear that post-project evaluation can help States and MPOs to better understand the costs and benefits of various project types, the costs associated with post-project evaluation are sometimes perceived as outweighing the benefits. However, CMAQ programs that are willing invest the resources needed to evaluate projects post-implementation and use the results to restructure and refocus their programs should be better able to meet their objectives moving forward.

Effective Practices from the Field

Evolution of the CMAQ Program Document

Denver Regional Council of Governments (DRCOG), Denver, Colorado

In 2008, DRCOG staff developed a 12-page public document that lays out, in an accessible, plain-English narrative, the evolution of the MPO's 16-year history of using CMAQ funds. It includes how the agency's approach and priorities for using these funds have changed over the years in response to the changes in planning, air quality, and financial contexts in the Denver metropolitan area. The document highlights examples of the types of programs and projects funded by CMAQ in the past and identifies current DRCOG priorities. The guide is an excellent resource for helping new or potential project sponsors understand the regional CMAQ process.

"After-Action" Debriefing and Restructuring of the CMAQ Evaluation Committee Membership

Southwestern Pennsylvania Commission (SPC), Pittsburgh, Pennsylvania

After it delivers a recommended suite of CMAQ projects to the SPC board for inclusion in the TIP, the CEC holds an "after-action" debriefing to document both positive and negative aspects of the recently completed 2-year CMAQ project selection process. At the end of the debriefing, the CEC is dissolved. At the beginning of each future CMAQ cycle, the membership of the CEC is revisited, and new members and organizations have the opportunity to participate. The results of previous debriefings are used to improve processes for the next CMAQ cycle and to educate new committee members about CEC goals and functions.

Streamlining of CMAQ Project Review Process

Regional Planning Commission of Greater Birmingham (RPCGB), Birmingham, Alabama

In evaluating its CMAQ program, RPCGB found that its CMAQ projects were taking too much time to be designed and built. The MPO felt that its smaller CMAQ projects were being subjected to the same design standards as larger projects. Working with the Alabama Department of Transportation (ALDOT), RPCGB developed procedures to speed up the project review process. This included shifting responsibility for planning, design, and the bid process to the project sponsor. This has resulted in a shorter design and bidding period and has reduced project costs for these smaller-scale projects.

Evolution and Integration of CMAQ Processes Result in Emergence of a "CMAQ Mentality"

Metropolitan Transportation Commission (MTC), San Francisco Bay Area, California

In San Francisco, the CMAQ is not operated as a stand-alone program but is viewed instead as just one funding source among many. As such, CMAQ funds are pooled with other funding sources in support of multimodal program areas supporting a broad range of project types. MPO staff observed the significance of CMAQ and its historical impact on regional policy development. The CMAQ program supported the agency's philosophical shift, over the past 2 decades, to consider improved air quality and quality of life as important factors in evaluating all transportation projects, not just those that are CMAQ-eligible. As a result, one staff member noted, MTC now conducts all programming with a "CMAQ mentality."

3. OBSERVATIONS FROM THE FIELD: BENEFITS, CHALLENGES, AND OPPORTUNITIES

All of the information presented in this report is intended to contribute to further discussions about the effectiveness of the CMAQ program, both as it currently stands and with an eye to the future. Rather than focusing solely on identifying effective practices themselves, some of the questions asked during site-visit interviews centered on better understanding the broader planning and funding context within which State DOTs and MPOs develop their CMAQ programs. This section presents the resulting observations in four categories:

- Links to the bigger picture.
- State DOT involvement.
- State air quality agency involvement.
- Data and analysis requirements.

Links to the Bigger Picture

The case studies indicated that CMAQ helps agencies to link more effectively to the "bigger picture" of regional planning by:

- Improving MPOs' planning capacity overall due to the oversight and operations required at the MPO level.
- Testing innovative projects on a small scale for demonstrable success before scaling up.
- Allowing broader policies to evolve that reflect the nexus between transportation and other issues.
- Opening the transportation planning and programming process to nontraditional partners.
- Building stronger relationships with and obtaining involvement of air quality agencies.

For many of the MPOs interviewed, the CMAQ program is one of the largest sources of Federal transportation discretionary funds available. As such, it is one of the few areas for which MPOs themselves, rather than State DOTs, take the lead in planning and programming. Direct decisionmaking authority for CMAQ has required MPO staff to engage their policy boards, subcommittees, and community stakeholders to develop methodologies to effectively evaluate and program CMAQ projects. The scrutiny to which CMAQ projects are subject in terms of eligibility requirements and calculating and reporting benefits forces MPOs to develop planning and programming processes that, in some ways, are more rigorous than for other Federal funding sources. This not only builds MPOs' internal capacity to program CMAQ funds but also improves metropolitan planning processes overall and positions MPOs to plan for other sources of transportation funding, whether local, State, or Federal.

Some MPOs explained that they have been able to use the CMAQ program as an incubator for testing new and innovative projects that might not otherwise qualify for Federal funds. Most agencies noted that they use CMAQ to support popular projects that might not be able to receive funding without it, such as diesel retrofits and transportation demand management (TDM) activities. For example, when CMAQ-funded projects have been able to demonstrate their value and success on a smaller scale, agencies have been able to justify greater resource allocation to them in subsequent programming rounds on a larger scale. The Birmingham MPO noted that "Birmingham is an example of a place where projects that were originally funded with CMAQ are now funded through other programs as well."

Though it is sometimes seen as a burden, the CMAQ program's additional requirement to estimate project benefits has a positive side in that it brings a higher degree of analytical rigor and justification into the planning and programming process. Both the Birmingham and Pittsburgh MPOs noted that projects passing the CMAQ eligibility test have an advantage in determining funding because they are better able to estimate their benefits than are most traditional projects, which are not subject to the same degree of analysis or evaluation. The CMAQ model of analysis may play an increasingly important role in determining funding in a future, more performance-based model of transportation planning.

Several of the larger MPOs interviewed reflected on the nearly 20-year history they now have of programming CMAQ funds over the years. These MPOs revealed how the CMAQ program's success on a smaller scale has led to shifts in their boards' policy goals at a broader level, enabling them to see the nexus between transportation planning and other quality-of-life issues ranging from public health to community livability and environmental stewardship.

The CMAQ program has also helped MPOs and State DOTs to expand their reach by opening the transportation planning and programming process to nontraditional partners such as trucking companies, school districts, and citizen groups. This builds awareness and understanding of the transportation planning process to a broader audience of stakeholders and also challenges MPOs to improve their community outreach and public relations. In Pittsburgh, the Downtown Business Association became a project sponsor and worked successfully with the MPO to start a pilot car-sharing program with seed funding from the CMAQ program.

State DOT Involvement

State DOTs play an important role in shaping CMAQ program development at the MPO level, taking different approaches in their type and level of involvement. The relationship between a State DOT and each of the MPOs eligible for CMAQ funding within that State is a critical factor in shaping how regional CMAQ programs function. State DOTs' approach to CMAQ influences the development of regional CMAQ programs in a number of ways:

- Each State DOT chooses how to allocate CMAQ funds to eligible MPO regions; some DOTs allocate funds to eligible MPO regions but also retain a portion of the CMAQ funds to run separate Statewide CMAQ program as well.

- Some State DOTs provide technical assistance or other extra support to MPOs to support improved decisionmaking (e.g., guidance on how to measure air quality impacts of hard to estimate projects).

- Some State DOTs have representatives sitting on the key committees that make decisions about CMAQ eligibility and project prioritization and selection.

State DOTs Suballocate CMAQ Funds to MPOs

The Federal-Aid Highway Program apportions CMAQ funds to States on a formula based on the size of nonattainment area populations and the severity of air quality deficiencies. Each State has the flexibility to suballocate funds among eligible MPO regions as it chooses. How States decide to suballocate funds among their MPOs can differ significantly from one State to another. Many States distribute all of the CMAQ funds directly to MPOs for programming. The allocation of CMAQ funds in Colorado, for example, is based on State legislation that requires 100 percent of funds to be given directly to MPOs for programming. In California, Caltrans (the State DOT) devolves programming control of nearly 100 percent of CMAQ funds directly to MPOs[16] around the State on the basis of a formula that accounts for population size and the severity of air quality nonattainment in their air basins; a small portion of the overall funds for the five non-MPO areas around the State are reserved for those that are in nonattainment and therefore are eligible to receive CMAQ dollars.

---

[16] Due to the large population of the Los Angeles MPO region, funds are suballocated directly to the region's County Transportation Commissions.

While many States treat CMAQ as an automatic pass-through that is distributed directly to MPOs, some take a more hands-on approach, retaining a pool of funds "off the top" for Statewide initiatives and splitting funds between regional and Statewide projects. For example, the Massachusetts Executive Office of Transportation (EOT) retains about $12.5 million a year for Statewide CMAQ projects and allocates the remaining funds directly to MPO regions. Maintaining a CMAQ program at the State level allows Massachusetts to achieve economies of scale and to take on projects that might be harder to develop if MPOs were working on them in isolation. For example, rather than distributing funds to the State's 13 MPOs for individual rideshare and other TDM programs, EOT uses State CMAQ money to fund the MassRides program, a one-stop shop for commuter information and ride-sharing across the State. EOT believes that MassRides creates Statewide brand recognition for rideshare and other commuter alternatives and that it is able to address complex commuting patterns, which are rarely confined to MPO planning boundaries, in a more cost-effective manner than MPO-specific programs would be capable of doing. Having a pool of CMAQ funds at the State level also provides EOT with the opportunity to collaborate with State-level air quality, public health, and environmental agencies, thereby building new relationships that can be leveraged in other aspects of planning and policy work and enabling innovative, nontraditional transportation projects, most of which would not qualify for State transportation funds.

> The Massachusetts Statewide CMAQ program includes the following programs:
>
> - School Bus Retrofit ($15 million in total)
> - MassRides ($2.5 million annually)
> - Enhanced Inspection and Maintenance (I&M)
> - Alternative Fuels for Statewide Fleets pilot program
> - Hybrid Vehicles Plug-in demonstration project
> - Safe Routes to School program

Whether the State DOT distributes funds directly to the MPOs by formula or retains some of the money for a Statewide program of its own, CMAQ funding suballocations can be contentious as MPOs vie with one another—and with the State itself, when a Statewide CMAQ program exists—for larger portions of the overall CMAQ allocation. When States revisit and update their allocation formulas, the funding impact is that some MPOs "win" and others "lose." For example, when Oregon DOT (ODOT) recently updated its regional CMAQ allocation formula; the Medford MPO went from receiving just 5.5 percent of CMAQ funds Statewide to 13 percent, more than doubling the size of its CMAQ program. Since 1991, the Birmingham MPO is the only area in Alabama that is eligible for CMAQ funding. But currently, the State anticipates three new areas in Alabama soon be redesignated as nonattainment. This would have a dramatic impact on how the CMAQ program is operated in Birmingham, since ALDOT would be suballocating CMAQ funds to four areas around the state instead of just one.

State DOTs Can Be a Source of Technical Support and Capacity-Building for MPOs

States can serve in an important advisory role on technical capacity issues for MPOs that are struggling to meet complex reporting requirements or develop effective methods for estimating emissions reductions across nontraditional transportation project proposals, especially for smaller and midsized MPOs. PennDOT has taken a leadership role in helping to streamline CMAQ project calculations across planning regions by hiring a consulting firm to develop standardized spreadsheets to estimate the emissions benefits for a variety of air-quality-type projects. This work has resulted in a stand-alone software package that small and large MPOs alike can use to evaluate potential projects. PennDOT keeps the consultant on contract to provide services on an as-needed basis, thus helping MPOs as challenges arise when new types of proposed CMAQ projects cannot be matched up with existing spreadsheets[17]. This approach saves MPOs the cost of reinventing the wheel each time a new methodology needs to be developed. It also ensures that calculations are more robust and consistent across Pennsylvania, which increases the meaningfulness and integrity of the results. It may also help less traditional projects to compete more effectively within the transportation funding process. For example, when the Pittsburgh Downtown Partnership (PDP) came to SPC, the Pittsburgh MPO, with a proposal to fund a demonstration

---

[17] PennDOT staff noted that this happens in about 10 percent of cases.

car-share program with CMAQ money, SPC was able to utilize the emissions spreadsheets developed by PennDOT's consultant for a car-share program that was already in existence in Philadelphia and to make a case for funding it. PennDOT also helps to build self-sufficiency for MPOs by convening a Statewide working group that meets quarterly to discuss issues and challenges faced by different MPOs statewide and discuss strategies for overcoming shared challenges.

State DOTs Sit on Committees Determining CMAQ Eligibility or Project Prioritization

States may play a role in determining project eligibility for CMAQ as well as in prioritizing and even selecting projects. Representatives of State DOTs commonly sit on MPO boards, which make the final determinations of which projects ultimately receive funding. Sometimes they also sit on the MPO subcommittees that oversee the technical analysis of projects and make recommendations to the MPO board on how projects should be prioritized and selected from the applicant pool. Some States also convene a Statewide interagency committee to make CMAQ eligibility determinations for proposed projects and to seek input on prioritization. (Sometimes it is the same group that meets to make conformity determinations.) In Massachusetts, EOT convenes a statewide CMAQ eligibility committee, including the Department of Environmental Protection (MassDEP), which meets annually to determine which project proposals qualify for CMAQ funding. In many cases, State DOTs are project sponsors for CMAQ projects, working with the MPO to program projects. PennDOT districts, for example, commonly submit CMAQ project proposals to the MPOs for funding consideration.

Air Quality Agency Involvement

The U.S. Code Title 23, Section 149, states *"the Secretary shall encourage States and Metropolitan Planning Organizations to consult with State and local air quality agencies in nonattainment and maintenance areas on the estimated emission reductions from proposed congestion mitigation and air quality improvement programs and projects."*[18] FHWA guidance echoes this, stating that State DOTs and MPOs should consult with State and local air quality agencies to develop and program projects that have the greatest impact on air quality.[19] There is no uniform National model for how air quality and transportation organizations should work together on the CMAQ program, however. Specific approaches and structures are determined locally, producing a variety of models of interaction. There are two main ways in which air quality agencies at the seven site locations are consulted on the CMAQ program:

- Air quality agencies serve as members on the policy, technical, or evaluation committees for MPOs.
- Air quality agencies propose CMAQ projects for future implementation.

Interviewees from the site visits noted that the CMAQ program has played a role by shifting this dynamic and providing an opportunity for air quality and other environmental agencies to build stronger working relationships with MPOs and State DOTs. Serving on CMAQ committees and/or proposing CMAQ projects enabled air quality and environmental agencies to engage in the transportation planning process as partners and participants. However, some interviewees acknowledged that barriers to full engagement and consideration of air-quality impacts still exist. If the role of the air quality agency is too narrowly defined, some important information about a project's emission reduction potential could be missed. For example, greater involvement by the air quality agency could help inform project selection by providing information about the relative magnitude of the emission reductions or the effectiveness of reducing a particular pollutant.

---

[18] 23 U.S.C. §149 *(g) Interagency Consultation;* available at http://www.law.cornell.edu/uscode/uscode23/usc_sec_23_00000149----000-.html
[19] Federal Highway Administration, *The Congestion Mitigation and Air Quality (CMAQ) Improvement Program under the Safe, Accountable, Flexible, Efficient Transportation Equity Act: A Legacy for Users, Final Program Guidance;* available at http://www.fhwa.dot.gov/environment/cmaq08gm.htm.

Air Quality Agencies Serve as Members of Transportation Boards

In each of the locations visited, the air quality agency was a member of a board or committee that dealt with the transportation and air quality interface. In the San Francisco Bay area, the MPO does not have a stand-alone CMAQ program. It instead integrates air quality issues throughout its planning process. Both the Bay Area Air Quality Management District (BAAQMD) and the San Francisco Bay Conservation and Development Commission (SFBCDC) serve as voting members of the MTC Joint Policy Committee, which sets the overall vision and policy for the MPO. Similarly, in Denver, the Colorado Regional Air Quality Council (RAQC) is a member of the Regional Transportation Committee, which provides overall policy direction for the MPO. More typically, the air quality agency is a member of a committee that reviews, evaluates, and selects projects for CMAQ funding. In Fort Collins, the CDPHE is a voting member of the CMAQ Evaluation Committee, which reviews and ranks all CMAQ proposals and then makes recommendations to the MPO policy board. Pittsburgh is similar to Fort Collins, with the Pennsylvania DEP and the Allegheny County Health Department sharing a role as a voting member of the CMAQ Evaluation Committee. In Massachusetts, CMAQ projects are not evaluated independently from the TIP evaluation process. Instead, the State has a CMAQ Eligibility Committee on which MassDEP is a voting member. The committee determines if a proposed project is eligible for consideration for CMAQ funding. All projects are then programmed in the TIP and approved by the MPO policy board.

Air Quality Agencies Serve as Project Sponsors

The CMAQ program has provided the air quality agencies with the opportunity to utilize CMAQ funds to help further their transportation-related air quality initiatives. In some cases, the air quality agency is able to utilize local funds or private investment to leverage Federal CMAQ dollars. The air quality agencies in each of the seven MPOs visited have sponsored CMAQ projects.

In Massachusetts, MassDEP utilizes CMAQ funding to modernize the Statewide Vehicle Inspection and Maintenance (I&M) program and the retrofitting of all eligible school buses, among other initiatives. In Pittsburgh, the Allegheny County Health Department serves as a project sponsor on behalf of local municipalities. During the last round of CMAQ funding, the Health Department used fines that it collected from stationary-source air quality violations within the region to pay the local match for a series of diesel retrofits for public works vehicles for a consortium of 12 municipalities. In Birmingham, the Alabama Department of Environmental Management (ADEM) is a member of the Clean Air Alliance, which includes private-sector firms. For the ADEM truck stop electrification project completed in 2006, CMAQ funds were only 11 percent of the total cost. The private sector contributed the remaining 89 percent. In Denver, the RAQC received more than $1.6 million over a 2-year period to support its ozone reduction program. In San Francisco, the BAAQMD used CMAQ funds to support its Spare the Air campaign, and in Medford, Oregon, the State's lead air quality agency, the Oregon Department of Environmental Quality (ODEQ), sponsored the region's first PPP, a diesel retrofit to allow Rogue Disposal & Recycling Company's fleet to burn cleaner fuels.

Data and Analysis Requirements

Many of the MPO and State DOT staff interviewed expressed support for the documentation of project benefits required for CMAQ eligibility because they felt it led to the selection of higher-quality projects overall. In Birmingham, for example, MPO staff expressed greater confidence in justifying CMAQ projects to the public than other Federally funded projects because of the analysis that CMAQ projects must withstand in order to be selected. A number of agency staff felt that the CMAQ program raises the bar for other projects competing in the Federal-Aid funding process and suggested that the program be used as a model for other Federal-Aid programs, noting that if all Federally funded projects had to meet the same standards as CMAQ, better projects would be built across the board.

At the same time, one MPO noted that the data requirements associated with determining CMAQ eligibility can add to project costs. For example, one agency noted that a major project such as a bridge rehabilitation using Federal funds requires less justification of benefits derived than does a relatively small bicycle project using CMAQ funds. Similarly, an intersection improvement project would require different levels of analysis depending upon whether CMAQ funding would be used to support the project or not.

In addition to the quantification of benefits, CMAQ funds are tied to other requirements that limit eligibility for project proposals. For example, Federal guidance recommends a 3-year cap on the use of CMAQ funds for operating expenses.[20] Agencies differ in how they think about additional requirements such as this. Some agencies felt that this places an unnecessary restriction on projects whose air quality benefits may improve over time; however, in Boston, this restriction was turned into an advantage by requiring an increasing local match each year for operations projects. The belief was that only projects that are increasingly self-reliant should be sustained with funding after Year 1 or 2.

Some agencies asked why additional requirements are applied to CMAQ when it is such a small percentage of the overall FHWA funding program, as this may cause agencies to devote a disproportionately high amount of staff time and resources to administer CMAQ funds and oversee projects in order to meet the requirements. As one staff member noted, CMAQ represents about 4 percent of the TIP program yet can consume about 20 percent of staff time. Since the CMAQ program itself does not include funding for agency administration and oversight, agencies must use their Federal planning funds to cover these costs.

Many agencies agreed that CMAQ's additional requirements can be seen as positive at a broad level since more analysis should lead to higher-quality projects being supported, but they noted that it was a challenge to allocate the increased level of staff time required for administering CMAQ to a funding source that represents such a small portion of their overall transportation budgets. Some staff noted that if increased planning funding were included to cover additional administrative, reporting, and evaluation costs, the additional CMAQ evaluation requirements might serve as an excellent model for how all Federal funding categories could be run more effectively. Under this scenario, the best projects will rise to the top under more rigorous analysis and with a clearer delineation of expected benefits.

---

[20] Federal Highway Administration, *The Congestion Mitigation and Air Quality (CMAQ) Improvement Program under the Safe, Accountable, Flexible, Efficient Transportation Equity Act: A Legacy for Users, Final Program Guidance*, p. 13; available at http://www.fhwa.dot.gov/environment/cmaq08gm.htm.

4. PUTTING IT ALL TOGETHER: LEGISLATION AND GUIDANCE TRANSLATE INTO PROJECTS

This section of the report provides a unique view into how MPOs and State DOTs translate SAFETEA-LU provisions and FHWA guidance into innovative and valuable projects in the field. The projects highlighted here represent efforts that the interviewed MPOs felt were successful in areas such as:

- Providing air quality benefits.
- Strengthening interagency cooperation.
- Increasing quality of life and economic vitality in the community.
- Leveraging funds to maximize impact across geographic boundaries.

These project descriptions point to the variety of CMAQ activity across the country and provide a closer examination into the range of processes, projects, relationships, and impacts the CMAQ program is generating at a small sampling of agencies Nationwide. Through these project descriptions, readers will gain insight into how SAEFTEA-LU requirements and FHWA guidance are being used by States and MPOs to solicit, evaluate, and program proposed CMAQ projects at the state and local level.

Birmingham Truck Stop-Electrification

Location: *Birmingham, Alabama*
Cost: $1.8 million total, $170,000 in CMAQ funds
Project sponsor: Alabama Partners for Clean Air (APCA)
Estimated annual emissions reduction benefits:

Pollutant	Est. Annual Emissions Reduction (kg/yr)
Ozone	148
Ozone NO_x idle emissions factor	2,920
PM 2.5 (standard PM)	80

Project Partners

The partnership comprised 11 organizations, including the Alabama Department of Environmental Management (ADEM), the Jefferson County Department of Health, and the Alabama Trucking Association.

CMAQ Application Process

APCA worked with public- and private-sector firms to package this proposal. The private sector agreed to provide the electrification location, and the public sector agreed to provide some start-up financial assistance. The project was then proposed to Regional Planning Commission of Greater Birmingham (RPCGB).

Project Overview

The purpose of this project is to improve air quality by reducing the exhaust from idling engines. An idling diesel engine consumes about one gallon of fuel per hour. The exhaust from the idling engines contains numerous pollutants, including nitrogen oxide (NO_x), carbon monoxide (CO), and volatile organic compounds (VOCs). Because truck drivers are required to get several hours of rest to ensure safe driving, they normally keep engines idling at truck stops during rest periods to provide heating and air conditioning in the cab and sleeping area as well as power for various appliances. The Truck Stop Electrification project is a partnership with a private-sector truck stop to install technology that allows truck drivers to shut off their engines while idling and to hook up a window adaptor that provides the truck with required heating and cooling, electricity, and communications services. The project includes the capital cost to install and operate the technology.

Estimation of Air Quality Benefits

Using Environmental Protection Agency emissions factors for calculating air quality benefits, it was estimated that the cost-effectiveness for reductions in NO_x was approximately $34 per kilogram per year.

MPO Project Evaluation and Selection Process

This project was part of an APCA package of proposals that was sent to the Metropolitan Planning Organization (MPO) Interagency Consultation Group to determine whether projects were eligible for CMAQ funding. Once this project was deemed eligible, the MPO Transportation Improvement Program (TIP) Subcommittee voted to allocate CMAQ funding to it.

Post-Project Evaluation

The project has been in operation since 2004, and the operator has collected data on how frequently the electrification technology is used. Based on the frequency of use data, APCA estimated that the project saved 178,000 gallons of diesel fuel in 2007 and reduced pollutants by 1,900 metric tons.

Conclusion

In addition to the air quality benefits, the project sponsor noted other benefits such as energy independence through fuel savings, economic development, and the provision of better working conditions for the truck drivers.

Statewide School Bus Retrofits in Massachusetts

Location: *Massachusetts Statewide*
Cost: $16.4 million in total CMAQ funding over 3 years
Project sponsor: Massachusetts Department of Environmental Protection (MassDEP) and Executive Office of Energy and Environmental Affairs (EOEEA)
Estimated annual emissions reduction benefits:

Pollutant	Est. Annual Emissions Reduction (kg/yr)
CO	602,573
HC*	30,129
PM	1,998

*HC = hydrocarbons.

Project Partners

Project partners were the Massachusetts Executive Office of Transportation (EOT), MassDEP, and the EOEEA.

CMAQ Application Process

Massachusetts' two State environmental agencies, EOEEA and MassDEP, submitted a proposal to the Massachusetts EOT for a Statewide program to retrofit the diesel engines in all school buses across the Commonwealth.

Project Overview

The project provides diesel oxidation catalysts (DOCs) and crankcase filters (CCFs) for approximately 7,800 school buses in Massachusetts and will result in all school buses in the Commonwealth being retrofitted by 2010. EOEEA sponsored this project because retrofitting school buses with new technologies that address both external and in-cabin exposure to pollutants will provide significant air quality benefits at a relatively low cost.

EOT has transferred the funds to MassDEP to manage. The Agency coordinates directly with each of the school districts and bus operators to complete the work.

Estimation of Air Quality Benefits

To calculate the air quality benefits, MassDEP used the Environmental Protection Agency diesel emissions quantifier model, which is available online at: http://cfpub.epa.gov/quantifier/view/. The model estimated that installing DOCs and CCFs on older school buses Statewide would reduce carbon monoxide (CO) and HC emissions by 80 percent, and particulate matter (PM) emissions by 25 percent. Annual emissions reductions in kilograms were calculated for each pollutant and a cost-effectiveness rate was then determined. The estimated cost to reduce a kilogram of CO is about $30, while HC reduction is estimated at $597 per kilogram.

Statewide Project Evaluation and Selection Process

This project was part of the Statewide CMAQ program overseen by the Massachusetts EOT. Upon receiving the submission from MassDEP, EOT included the proposal on the agenda for the CMAQ Consultation Committee to determine its eligibility. After the project was recommended for funding, the committee referred the project back to EOT, for inclusion on the State Transportation Improvement Program (STIP). There is no formal evaluation committee to review Statewide CMAQ projects.

Post-Project Evaluation

A post-project evaluation has not been conducted for this program because it is still in its implementation phase.

Conclusion

EOT believes that this project is a good example of the benefits of having a Statewide CMAQ program that can promote broader programs and projects that generate benefits across multiple Metropolitan Planning Organization (MPO) boundaries. As a result of the project, diesel engines in school bus fleets will be retrofit throughout the entire state, resulting in substantial air quality benefits at relatively low cost.

Denver's Traffic Signal System Improvement Program (TSSIP)

Location: *Denver, Colorado*
Cost: $3.9 million (in FY08) in CMAQ funding
Project sponsors: Denver Regional Council of Governments (DRCOG)
Estimated annual emissions reduction benefits:

Pollutant	Est. Annual Emissions Reduction (kg/yr)
CO	656,932

Project Partners

Project partners included DRCOG, Colorado Department of Transportation (CDOT), local jurisdictions that operate traffic signals, and CDOT Regions 1, 4, and 6 (in all, 32 stakeholder entities).

CMAQ Application Process

TSSIP is one of six programs that now receive dedicated CMAQ funds, due to their long history and demonstrated success at meeting CMAQ goals in the Denver region. At the beginning of each Transportation Improvement Program (TIP) update cycle, DRCOG determines the specific funding commitments to these six programs. DRCOG updates the TSSIP generally on a 4-year cycle, programming 6 years of funding at a time.

Project Overview

The overall purpose of TSSIP is to reduce travel time and vehicle emissions within the DRCOG Transportation Management Area (TMA). The program involves both capital and traffic signal timing improvements on a regional level. The capital improvements, which are prioritized according to a FY 2008–2012 schedule, involve providing equipment and installing communications links to facilitate traffic signal timing. The traffic signal timing component coordinates more than 3,500 traffic signals across 32 jurisdictional boundaries and on all major roads in the DRCOG TMA. Signals are generally coordinated via time-of-day (TOD) plans such as "morning peak," "off-peak," and "afternoon peak." The TOD plans are updated on a 3- to 5-year basis for each corridor. TSSIP was implemented in 1994, programming an initial 6 years of funding, and was updated in 1996, 1999, 2003, and 2007.

Estimation of Air Quality Benefits

The air quality benefits are assessed on the basis of travel times and the number of stops measured in the corridors where timing was revised. Benefits from the capital improvement component of the program are reflected by subsequent corridor traffic signal timing improvements. With each signal retiming project, TSSIP expects to reduce corridor travel time by 5 to 15 percent. For 2007, TSSIP reported a savings of approximately 2.2 million vehicle-hours traveled and 1.4 million kilograms of carbon monoxide (CO) emissions.

MPO Project Evaluation and Selection Process

After a call for projects, DRCOG staff worked extensively with regional traffic signal operating agencies to verify current conditions, identify critical needs, and program upcoming funding. A copy of the summary report for the most recent funding cycle may be found at
http://www.drcog.org/documents/TSSIP%202007%20Summary%20Report.pdf.

Post-Project Evaluation

DRCOG conducts a post-project evaluation for every TSSIP project. The results of each evaluation are summarized in a project brief that is distributed to interested parties in the region. Annually, DRCOG summarizes the results of all completed projects for the *CMAQ Reporter*.

Conclusion

DRCOG believes that the CMAQ program has allowed the Agency to fund more air-quality-focused projects, including TSSIP, and has strengthened its working relationship with Regional Air Quality Council (Colorado). DRCOG also believes that the CMAQ program has allowed the Agency to focus explicitly on how transportation affects air quality in the Denver region.

Fort Collins Community Bicycle Library

Location: *Fort Collins, Colorado*
Cost: $132,000 in CMAQ funding (with local match of $33,000)
Project sponsor: City of Fort Collins
Estimated annual emissions reduction benefits:

Pollutant	Est. Annual Emissions Reduction (kg/yr)
CO	759

Project Partners

Project partners included Colorado State University, Downtown Development Authority, Bike Fort Collins (BFC), Fort Collins Bicycle Collective, Fort Collins Convention and Visitors Bureau, and local hotels and businesses.

CMAQ Application Process

Responding to a public solicitation by North Front Range Metropolitan Planning Organization (NFRMPO) for CMAQ projects, the City of Fort Collins applied for funding for the Community Bicycle Library project. The proposal was a joint effort by a coalition of partners that included a local university, downtown business groups, and bicycle organizations.

Project Overview

The Fort Collins Community Bicycle Library is a 2-year project (2007–2009) to lend bicycles to the public at no cost. The purpose of the project is to reduce vehicle miles traveled (VMT) and traffic congestion as well as to improve air quality. BFC, a local bicycle advocacy group, administers the Bike Library. Users can borrow one of 100 bikes from two locations in the City of Fort Collins for periods ranging from an hour to a week. The program's Web site, http://www.fcbikelibrary.org, allows users to reserve a bike for future use or to find out if bicycles are available. Web-based tracking allows the project sponsors to document usage. Within the first month of opening, the library was lending bicycles at capacity.

Estimation of Air Quality Benefits

The City estimated that the project would reduce carbon monoxide (CO) by 759 kilograms in the first year of operation. To quantify the air quality benefits of the Bicycle Library, the City worked with the MPO's CMAQ consultant to determine appropriate inputs and assumptions. This required some original research since the project was the first of its type for the MPO and the State.

MPO Project Evaluation and Selection Process

Using the information submitted by the City, the MPO's CMAQ Selection Committee assigned points for the project on the basis of both quantitative and qualitative measures. This CMAQ proposal was then ranked, along with others, by the Selection Committee, and a recommended list of projects for funding was developed. The MPO makes the final determination of projects to fund.

Post-Project Evaluation

To evaluate the success of the Bicycle Library, the City of Fort Collins developed a number of metrics and plans to conduct user surveys and collect Web-based data throughout the project to quantify VMT reduction, traffic congestion, and air quality improvements.

Conclusion

The City of Fort Collins and NFRMPO view the CMAQ program as a major success in furthering the goal of broadening the types of projects that are funded. City officials believe that the library is a model for interagency cooperation and that it has already had positive regional and social impacts beyond air quality benefits. It also has furthered the MPO's goal to reach out to nontraditional partners.

Medford, Oregon Diesel Retrofits

Location: *Medford, Oregon*
Cost: $50,000 in CMAQ funding (with $50,000 match by Rogue Disposal & Recycling)
Project sponsor: Oregon Department of Environmental Quality (Oregon DEQ)
Estimated annual emissions reduction benefits:

Pollutant	Est. Annual Emissions Reduction (kg/yr)
PM 10	6,402

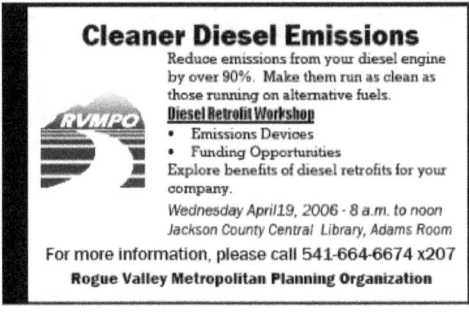

Project Partners

Project partners included Rogue Disposal & Recycling, Oregon DEQ, and Rogue Valley Council of Governments (RVCOG).

CMAQ Application Process

Rogue Disposal & Recycling, a private waste management and recycling company, learned about the opportunity to apply for CMAQ funds at one of two open houses on diesel retrofits organized by RVCOG staff. Oregon DEQ submitted an application on behalf of the company because only public agencies are eligible to apply for Federal funds from Rogue Valley Metropolitan Planning Organization (RVMPO).

Project Overview

The project goal was to reduce cancer and asthma risk from exposure to diesel emissions by installing emissions control equipment on selected vehicles in the Rogue Disposal & Recycling fleet.

Estimation of Air Quality Benefits

Oregon DEQ staff developed formulas to estimate the air quality benefits of Rogue Disposal & Recycling's diesel retrofits, which cut emissions on each vehicle by 50 percent. Before the retrofit, each vehicle ran on low-sulfur diesel (500 parts per million [ppm]); after the retrofit, each could run on ultra-low-sulfur diesel (15 ppm).

MPO Project Evaluation and Selection Process

At first, RVMPO's Technical Advisory Committee (TAC) was reluctant to support the project because CMAQ funds had never been given to a private company or used to fund diesel retrofits in the MPO region before. However, Oregon DEQ analysis showed that the project would substantially reduce vehicle emissions, and additional analysis showed that diesel retrofits are more cost-effective, per ton equivalent of air pollution removed, than nearly any other CMAQ-eligible project type. Those analyses provided the foundation for TAC's ultimate recommendation to find the project.

Post-Project Evaluation

RVMPO does not conduct post-project evaluations for every CMAQ-funded project, but it does require evaluation of diesel retrofits to ensure that air quality targets are met before granting subsequent funding.

Conclusion

After successfully completing this project, the RVMPO Policy Committee sees diesel retrofits as an important and cost-effective component of the CMAQ program. The 2008–2011 TIP obligates $750,800 for five diesel retrofit projects estimated to reduce annual PM 10 emissions by 80.7 tons.

Pittsburgh Downtown Car-Sharing Program

Location: *Pittsburgh, Pennsylvania*
Cost: $200,000 in CMAQ funding (with $50,000 local match from the Pittsburgh Downtown Partnership)
Project sponsor: Pittsburg Downtown Partnership (PDP)
Estimated annual emissions reduction benefits:

Pollutant	Est. Annual Emissions Reduction (kg/yr)
CO	4,212
NO_x	456
PM 2.5	7

CMAQ Application Process

PDP submitted a proposal for $200,000 of CMAQ funding during Southwestern Pennsylvania Commission's (SPC) 2005 call for CMAQ projects, with a $50,000 match from PDP. The application was to provide financial support over a 2-year period to get the program started.

Project Overview

PDP wanted to attract car-sharing to downtown Pittsburgh to provide a transportation alternative for residents as well as employers and employees in the area. The idea arose in response to a growing job market and renewed interest in downtown living. Private car-sharing companies had launched successful programs in large cities like New York and San Francisco but were skeptical about the market potential of a midsized city like Pittsburgh. In order to entice private car-sharing companies to invest in the Pittsburgh market, the PDP's CMAQ proposal included funding for a feasibility study, marketing and outreach to expand membership and participation in the program, and revenue guarantees to the private-sector vendor in the initial months of operation until the program became self-sustaining.

Estimation of Air Quality Benefits

It was estimated that the project would reduce carbon monoxide (CO) by 1.54 kilograms per day, nitrogen oxide (NO_x) by 1.25 kilograms per day, and fine particulate matter (PM 2.5) by 0.02 kilogram per day. CMAQ had never been used to fund a car-sharing program in Pittsburgh before, so SPC did not have any standard assumptions or calculations for determining air quality benefits. Using membership and follow-up survey data from Philly CarShare, which had been in operation for several years at that point, Pennsylvania Department of Transportation's on-call CMAQ consultant developed assumptions about travel behavior change to calculate the air quality impacts of car-sharing and establish the project's funding eligibility.

MPO Project Evaluation and Selection Process

After refining the calculation of air quality benefits and the cost-effectiveness measures, the project was evaluated by the SPC CMAQ Evaluation Committee. The committee reviewed the project assumptions and voted to include the project among the recommended set of CMAQ projects in the 2007–2010 Transportation Improvement Program, which was adopted by SPC in June 2006.

Post-Project Evaluation

CMAQ funds were used over a 2-year period during start-up of the Pittsburgh car-sharing program. During that period, PDP tracked program usage and provided regular reports to SPC. The pilot was successful. The Pittsburgh car-share is now self-supporting and is slowly growing and expanding on its own.

Conclusion

PDP believes that there are desirable outcomes beyond the air quality benefits for the project. The additional benefits include positively affecting the economic vitality of the downtown area and providing employees and residents with alternative travel options.

Bay Area Freeway Service Patrol (FSP)

Location: *San Francisco, California*
Cost: $5 million (CMAQ funds used for an average of approximately 10 percent of total cost)
Project sponsors: MTC Service Authority for Freeways and Expressways (MTC SAFE), California Highway Patrol (CHP), California DOT (Caltrans)
Estimated annual emissions reduction benefits:

Pollutant	Est. Annual Emissions Reduction (kg/yr)
CO	95,000
NO_x	445,000

Project Partners

Project partners included CHP, MTC SAFE, Caltrans, and private tow contractors in the region.

CMAQ Application Process

MTC SAFE does not apply for CMAQ funds through a standard fund application process; rather, funding for this program is supported by the operational strategy goals paid out in the Regional Transportation Plan (RTP). Specific funding levels are established cooperatively with MTC's transportation partners while considering how to strike the best balance among programs.

Project Overview

The purpose of FSP is to improve freeway safety, reduce congestion caused by vehicle accidents and incidents, and enhance air quality by reducing stop-and-go traffic. MTC SAFE primarily administers FSP, while CHP has field operation and supervision responsibilities. Caltrans provides data collection and route assessment support. A fleet of 85 specialized, state-of-the-art tow trucks patrol 35 routes stretching 540 miles on the Bay Area's freeways, and offer assistance to stranded motorists. Tow-truck drivers also clear road debris to keep traffic moving. The tow trucks primarily operate during the morning and afternoon weekday commuting hours. The Bay Area FSP, which began in 1992, was modeled on a program that started in Los Angeles in July 1991. On average, FSP drivers annually assist approximately 670,000 vehicles.

Estimation of Air Quality Benefits

The project was estimated to reduce 19 kilograms per year of reactive organic gases, 445 kilograms per year of nitrogen oxide (NO_x) and 95 kilograms per year of carbon monoxide (CO). MTC SAFE developed a benefit-cost model to assess cost, fuel, and air quality benefits of FSP. It estimated that FSP provides 2,278 hours of delay savings and 5,818 gallons of fuel savings, for a benefit-cost ratio of 3.6:1.

MPO Project Evaluation and Selection Process

MTC SAFE does not use a uniform scoring method to select projects for CMAQ funding. The primary criterion is the extent to which projects meet goals in the long-range RTP. RTP program area contacts may develop additional criteria or methods to identify CMAQ-eligible projects.

Post-Project Evaluation

MTC SAFE conducts post-project evaluations for selected projects, including FSP. A performance evaluation conducted for FSP indicated that the program was very cost-effective. This finding provided a basis of support for MTC SAFE to continue funding and expanding the program.

Conclusion

MTC SAFE believes that the FSP program is a cost-effective strategy to meet regional planning goals of improving safety and air quality while reducing congestion. Due to the ongoing benefits of the program, MTC SAFE believes it would continue to fund FSP even if CMAQ funds were unavailable.

5. Conclusions

Phase II of the CMAQ Evaluation and Assessment Study was conducted in response to SAFETEA-LU requirements. It examined MPO and State DOT practices in seven locations Nationwide to gain a better understanding of the range of approaches that diverse agencies take in planning, programming, and evaluating CMAQ funds. It also sought to find, document, and share effective CMAQ implementation practices used by States and MPOs with other agencies around the country. Specific examples of effective practices were identified and the benefits and challenges of CMAQ implementation at the State and local level were explored. Section 4 highlights individual projects that the interviewed agencies felt were particularly successful at meeting their regions' transportation and CMAQ goals. Section 6 provides a brief overview of each of the seven day-long interviews conducted by the Federal CMAQ Team to prepare this report. Appendix A lists points of contact for each of the site visits to allow readers to contact case study representatives and obtain additional information.

While the CMAQ program is Federally funded, no National standard or set of regulations exists for how a CMAQ program should be structured and operated at the State or MPO level. It is intentionally left to the State or MPO to develop and operate a program that best responds to local and regional needs. Reflecting this, the Federal CMAQ Team found differences in structure and operations at each of the sites visited.

The seven case study locations visited in Phase II were selected based on the results of the Phase I CMAQ Evaluation and Assessment Report. Although they represent a very small sample of the total number of CMAQ programs being implemented across the country, the information gathered during the day-long interviews allowed the Federal CMAQ Team to draw several conclusions about potential effective practices as well as benefits, challenges, and opportunities presented by the CMAQ program nationwide.

Making project solicitation, prioritization, and selection processes more transparent plays an important role in engaging citizens and increasing stakeholder involvement.

An effective and transparent process helps to ensure that the CMAQ program's goals, policies, and procedures are clear and understandable to both the public and to potential project sponsors. In today's Web-based world, many MPOs find it effective and efficient to post the program description and project application online to allow wide and easy access to important program information.

An open and accessible CMAQ process provides citizen groups with a good introduction to the transportation planning process because many CMAQ projects deal with quality-of-life issues that these groups work on. As such, it provides a first step for school districts, community organizations, and private firms to learn about and participate in the metropolitan transportation planning process.

Standardizing processes to evaluate and rank multiple project types can be challenging, especially due to CMAQ's emphasis on nontraditional projects, but it is important for increasing transparency and gaining public confidence.

Proposed projects must estimate air quality benefits to be eligible for Federal CMAQ funding. Measures used to evaluate and rank projects are both quantitative and qualitative in nature.

Quantitative measures, which include a calculation of benefits, costs, and cost-effectiveness, are powerful because they help MPOs to better explain to the public how projects are ranked and funded, reinforcing the transparency of the selection process. In addition, quantitative measures allow MPOs to make decisions that are less subject to political whims and are easier for policy boards to justify. However, the way these measures are calculated and used to inform the project selection process may limit the effectiveness of the overall CMAQ program. Because FHWA guidance requires estimates only of short-term benefits (i.e., in the first year of project implementation), it is not clear that the best projects rise to the top. Estimating the impacts of projects over the long term requires forecast projections about travel activity, participation or usage rates, and other important factors farther into the future. While longer-term estimates may require more effort, they can improve the accuracy of emissions reductions estimates over time. This is an issue that several of the MPOs interviewed struggled with. Some MPOs are moving toward calculating a project's life-cycle costs and benefits in addition to the FHWA-required calculation of estimated benefits in the first year of project implementation. While the calculation of life-cycle benefits requires additional staff work, these MPOs feel it may provide a better overall estimation of a project's future cost-effectiveness. In addition, because of the technical nature of this work, many of the MPOs relied on outside support from either the State DOT or an outside air quality consultant for the development and refinement of the assumptions, equations, and calculations used in determining a project's eligibility and cost-effectiveness.

Qualitative measures normally supplement quantitative measures and allow a State or MPO to give priority to the types of projects that complement the agency's other goals and policies in addition to those pertaining to air quality and congestion. Examples of types of programs given priority include those within an urban development boundary, diesel retrofits, TDM, and bicycle and pedestrian programs.

When projects are subjected to both quantitative and qualitative screening based on clearly defined criteria and using models and assumptions understood by project sponsors, MPOs are better able to produce project rankings that are documented and understood by those interested in the process, thereby increasing public confidence in the transportation planning process overall.

Effective processes for evaluating program impact and ability to adapt in response to evaluations and/or changing conditions are important for continual program improvement.

An effective process for evaluating the CMAQ program's impact includes examining the overall CMAQ program as well as specific projects. MPOs that have the most effective CMAQ programs have adapted and modified their approach to programming CMAQ funds over the past 2 decades. Several of the MPOs discussed the importance of periodically evaluating the program overall to ensure its continued effectiveness in light of shifting regional conditions and evolving policy goals.

Competition for limited CMAQ funds has increased the need to conduct evaluations of specific projects. Several MPOs noted that post-project evaluation of CMAQ projects is being driven by CMAQ project sponsors.

While it is clear that post-project evaluation can help States and MPOs to better understand the costs and benefits of a project, the costs associated with the activity are sometimes perceived to outweigh the benefits. However, the most successful CMAQ programs are those willing to reevaluate their operations and programming and to use that information to restructure and refocus the program moving forward.

State DOTs play an important role in shaping CMAQ program development at the MPO level.

The relationship between the State DOT and each of the MPOs eligible for CMAQ funding within that State is a critical factor in shaping how regional CMAQ programs function. State DOTs influence the process in the following ways:

- State DOTs choose how to allocate CMAQ funds to eligible MPO regions; some DOTs allocate funds to eligible MPO regions but also retain a portion of the CMAQ funds to run separate Statewide CMAQ program as well.

- Some State DOTs provide technical assistance or other extra support to MPOs to support improved decisionmaking (e.g., guidance on how to measure air quality impacts of hard to estimate projects).
- Some State DOTs have representatives sitting on the key committees that make decisions about CMAQ eligibility and project prioritization and selection.

Some States play an important advisory role on technical capacity issues for MPOs that are struggling to develop effective methods for estimating emissions reductions across nontraditional transportation project proposals. States taking a more hands-on approach in CMAQ by reserving some funding for statewide initiatives have been successful in leveraging CMAQ funds on significant programs across regional boundaries. A number of examples in Section 3 indicate how State involvement can positively impact CMAQ implementation.

CMAQ engages air quality agencies as valuable partners in the transportation planning process.

The CMAQ program requires that State DOTs and MPOs work with State and local air quality agencies to develop and program CMAQ projects that have a positive impact on air quality. Because there is no single National model or standard, the approach and structure of these relationships is determined locally, producing a variety of models for interaction. At the seven site locations in this study, air quality agencies consult with State DOTs and MPOs during the CMAQ process in two major ways:

- Air quality agencies serve as members on the policy, technical, or evaluation committees for MPOs.
- Air quality agencies propose CMAQ projects for future implementation.

Section 3 provides insight into the range of positive impacts that CMAQ has had on engaging air quality agencies. Interviewees from the site visits noted that the CMAQ program has played an important role by providing an opportunity for air quality and other environmental agencies to build stronger working relationships with MPOs and State DOTs. Serving on CMAQ committees and/or proposing CMAQ projects has enabled air quality and environmental agencies to engage in the transportation planning process as partners and participants.

CMAQ's data and analysis requirements present both significant challenges and opportunities.

The CMAQ program requires documentation of a project's air quality benefits for funding eligibility. Overall, a number of agencies felt that CMAQ's additional requirements can be seen as positive at a broad level since an analysis should lead to the selection of higher-quality projects. There was a general recognition that meeting the minimum emissions analysis required by FHWA guidance may not produce an accurate estimate of the benefits of projects. However, the agencies noted that it was a challenge to allocate the increased level of staff time required for administering CMAQ to a funding source that represents such a small portion of their overall transportation budgets. Some staff noted that if increased planning funding were included to cover additional administrative, reporting, and evaluation costs, the additional CMAQ evaluation requirements could serve as an excellent model for how all Federal funding categories could be more effectively run. Under this scenario, the best projects will rise to the top due to more rigorous analysis and a clearer delineation of expected benefits.

CMAQ legislation and guidance are translated and applied to meet local transportation and air quality requirements and concerns, resulting in a range of innovative projects.

The projects highlighted in Section 4 of the report bring to life the variety of approaches used to select, program, implement, and evaluate CMAQ projects. The application of Federal CMAQ guidance responds to local conditions and transportation planning processes. As a result, the projects selected and implemented are finely attuned to an area's needs. The elements of effective CMAQ implementation outlined in Section 2, and the benefits, challenges, and opportunities documented in Section 3, are illustrated within the seven project summaries.

Positive impacts go beyond stated program goals.

An important finding of the Phase II study was that there are ancillary benefits derived from the CMAQ program in addition to the air quality and congestion benefit goals of the program. Nationwide, CMAQ funding represented 4.3 percent of FHWA's total authorizations between FYs 2005 and 2009[21] and at the seven MPOs visited, CMAQ accounted for just 2 to 4 percent of total Federal funds. Yet for a program representing such a small percentage of Federal transportation funding, site interviews suggested that CMAQ has a relatively large impact on broader transportation planning processes in MPO regions. The case studies indicated that CMAQ funds helped agencies to link more effectively to the bigger picture of regional transportation planning by:

- Improving MPOs' overall planning capacity due to the oversight and operations required at the MPO level.
- Testing innovative projects on a small scale for demonstrable success before scaling up.
- Allowing broader policies to evolve that reflect the nexus between transportation and other issues.
- Opening the transportation planning and programming process to nontraditional partners such as trucking companies, school districts, and air quality organizations.
- Building stronger relationships with and obtaining involvement of air quality agencies.

---

[21] *Safe, Accountable, Flexible, Efficient Transportation Equity Act: A Legacy for Users (SAFETEA-LU),* Funding Tables, August 2005; available at http://www.fhwa.dot.gov/safetealu/fundtables.htm.

6. SITE-VISIT CASE STUDIES

This section provides a brief overview of each of the seven locations visited as part of the Phase II CMAQ Evaluation and Assessment Project. Full-day interviews were held at each location with State, regional, and local officials involved in the planning, programming, and evaluation of CMAQ-funded projects.

Birmingham, Alabama: Regional Planning Commission of Greater Birmingham (RPCGB)

Introduction

The Birmingham MPO is responsible for comprehensive transportation planning in Jefferson and Shelby Counties in Alabama. Members of the MPO include local and State government officials as well as representatives from the Birmingham-Jefferson County Transit Authority and ALDOT. The two counties had a population of 835,000 in 2006. The Birmingham nonattainment area, comprising Jefferson and Shelby Counties, was originally classified as nonattainment for the 1-hour ozone standard by EPA on March 3, 1978 (43 FR 8962).[22] The nonattainment area at the time of initial classification was geographically defined as Jefferson County but was later expanded to include Shelby County. The MPO attained the 1-hour ozone standard and was redesignated as attainment on April 12, 2004. On June 15, 2004, Jefferson and Shelby Counties were classified as nonattainment for the 8-hour ozone standard. The area was redesignated to attainment for the 8-hour ozone standard effective June 12, 2006. The area designated as nonattainment for PM 2.5 includes all of Jefferson and Shelby Counties as well as a small portion of southern Walker County.

Since 1991, the Birmingham MPO has been the only area in the State of Alabama that qualifies for CMAQ funding. Currently, the State believes that three new areas may be redesignated as nonattainment in the near future. This would have an impact on how the CMAQ program is operated within Alabama since ALDOT has never had to suballocate CMAQ funds to any agency other than the Birmingham MPO.

CMAQ Objectives and Procedures

The CMAQ program operates within the confines of the overall TIP process. All of the projects or programs in the 4-year regional TIP support the overall goals and objectives in the MPO's long-range plan. In the most recent TIP, CMAQ accounted for about 4 percent of the total dollars programmed. The MPO evaluates all projects together and applies rankings, after which it determines the funding source. Every project competes against the others regardless of funding source. The MPO views this integrated approach as a strength. In the past, the MPO has set aside a small portion of funds for certain project types that it is promoting, such as bicycle and pedestrian projects and transit.

During the early years of the CMAQ program, most of the CMAQ funds were used for Intelligent Transportation Systems (ITS) and signalization-type projects. In the late 1990s, RPCGB decided to shift its priority to focus CMAQ on air quality concerns. The State has intentionally not included any transportation control measures (TCMs) in the State Implementation Plan (SIP). Working with ADEM, RPCBG began using the CMAQ program to include voluntary measures and projects to help achieve emissions reductions.

Using lessons it had learned from its FHWA Transportation Enhancements (TE) Program, RPCGB streamlined the procedures for the CMAQ program. Working with ALDOT, RPCGB developed procedures to speed up the project review process. This included giving the project sponsor responsibility for the planning, design, and bid process. The MPO felt that certain TE and CMAQ projects were being overdesigned. Small projects were being designed to highway standards, leading to higher project costs.

---

[22] Federal Register 43:8962, March 3, 1978.

Giving responsibility to the project sponsor can speed up the process and reduce project costs, but it requires monitoring by the MPO.

Project Selection and Funding

Each project under consideration for inclusion in the TIP is ranked, and the most appropriate funding source is then identified on the basis of funding eligibility. As one member stated, "A CMAQ project may not start and end the process as a CMAQ project. How it is funded can depend on available funding, and benefit analysis can shift a project between funding sources." The Birmingham MPO often mixes and matches CMAQ and attributable STP funds when deciding how to fund a project.

The MPO has a TIP Subcommittee comprising major project sponsors: the Cities of Hoover and Birmingham, the Birmingham-Jefferson County Transit Authority (BJCTA), ALDOT, and Jefferson and Shelby Counties. This subcommittee reviews the list of projects to ensure a balance by type, geography, and funding sources. For those projects that might be CMAQ-eligible, the MPO Interagency Consultation Group, comprising EPA, FHWA, FTA, ADEM, and the Jefferson County Health Department, among others, conducts a consultation process to determine the eligibility of proposed CMAQ projects.

One recipient of CMAQ funds is the Alabama Partners for Clean Air (APCA), an affiliation of 14 public, private, and nonprofit agencies that work with the MPO to implement voluntary strategies to improve air quality in the MPO. This agency provides public outreach and education about air quality issues, ride-sharing programs, vehicle fleet conversions, and emissions reduction efforts. It has received about $1.4 million in annual CMAQ funds for the past several years, which it uses to support a variety of air quality improvement projects.

To support the transit system, the region transfers about $3.2 million of CMAQ funds per year to transit. These funds are mostly dedicated to supporting public- and nonprofit-operated paratransit services. However, funding is also provided to BJCTA to support regular fixed-route transit services through the purchase of transit vehicles and the provision of maintenance support. CMAQ funds have also been used in the past, and are expected to be used in the future, to help jump-start new fixed-route transit, in particular suburban-to-urban commuter bus services. Alabama is one of only six States where the use of State gas tax funds for transit is prohibited.

Qualitative and Quantitative Measures

In order to standardize project assumptions and determine project effectiveness, the MPO commissioned a consulting firm to develop guidelines for air quality project effectiveness. The resulting set of assumptions and methodology covers most project types. If a project falls outside of the methodology, the MPO relies on information from other MPOs that might have evaluated a similar project. A project sponsor is responsible for providing information about a project, but the MPO staff usually apply the appropriate assumptions and then calculates the air quality benefits.

For the ozone awareness program, APCA has hired a market research firm to conduct public opinion surveys. One survey asked what actions people took in response to an ozone alert day. This helps the MPO to calculate benefits, gauge the overall effectiveness of the program, and refine the program for the future. One interviewee noted that the market research money is some of the best money the MPO spends as it gives important feedback on the program.

Reporting/Post-Project Evaluation

Because the Birmingham MPO is the only user of CMAQ funds within the State, the MPO prepares the documentation of CMAQ projects and benefits and forwards the information to ALDOT. ALDOT then transmits the information to FHWA for use in the CMAQ database.

Each of the projects that are part of the APCA program is required to provide pre- and post-project benefit reports. This allows the MPO to evaluate APCA projects and adjust funding for those types of projects that do not provide enough measurable benefits.

Boston, Massachusetts: Boston MPO

Introduction

The Boston MPO consists of 101 municipalities in Eastern Massachusetts. It is governed by a 14-member MPO board comprising the City of Boston (a permanent member); three cities; three towns, which rotate among the many municipalities in the MPO; and a mix of regional and State agencies. The entire State of Massachusetts is classified as moderate nonattainment for the 8-hour ozone standard. In addition, Boston and eight surrounding cities have a CO maintenance plan. Massachusetts Executive Office of Transportation (EOT) receives the Federal allocations for the State and distributes most of these funds directly to MPOs for programming on the basis of their population, but reserves a portion of the CMAQ allocation for projects with "Statewide significance." With the money the State retains, it has developed a substantial Statewide CMAQ portfolio, which includes projects like the Statewide school bus diesel retrofit program.

For FY 2008, the State and the Boston MPO programmed a total of $60 million in CMAQ funds. Major projects included $12.5 million for the Statewide school bus retrofit program, $6.2 million for the Statewide ITS, and $1.9 million programmed by the Boston MPO for transit hybrid locomotive switchers. Other project categories include signal timing adjustments, bicycle/pedestrian improvements, and the suburban mobility/TDM program.

CMAQ Objectives and Procedures

Many States treat CMAQ as a pass-through distributed directly to MPOs, but Massachusetts takes a more hands-on approach and allocates funds between regional and Statewide projects. The Statewide program is used to fund initiatives that span the various MPOs and benefit from having a single program. State officials feel that the flexibility they have to use CMAQ money for Statewide projects has been a success. Some programs, such as the Statewide park-and-ride initiative, are needed on a scale requiring a degree of cooperation between regional planning agencies that can be difficult to achieve. State staff feel that they can usher through some projects more expeditiously at the State level than would be possible at the regional or local level due to challenges that would be posed by the fragmentation of local governance.

The Boston MPO uses CMAQ as a funding mechanism but does not run a stand-alone CMAQ program. The agency seeks to address the overall goals of reduced congestion and improved air quality throughout its entire planning work, not solely through projects funded with CMAQ dollars.

The primary goal of the Boston MPO in programming CMAQ funds is to support the best overall projects with air quality benefits. Staff reported that having CMAQ as a funding source allows the MPO to undertake projects that might have had more difficulty competing if "transportation" were the only criterion at play in the evaluation. The school bus retrofit and truck stop electrification projects, for example, might not exist without CMAQ because most people would see them as "air quality" projects more than as "transportation" projects. A strong bicycle and pedestrian advocacy community exists in the Boston MPO, so the MPO often uses CMAQ in support of bicycle and pedestrian projects.

Project Selection and Funding

For at least a decade, the Boston MPO and the State as a whole tended to underfund the CMAQ program due to the high level of Federal funds dedicated to Central Artery projects. This has now changed, and the State has a target of programming 100 percent of available CMAQ funds. Beginning in 2005, the State EOT began setting CMAQ programming "targets" for each MPO to help ensure that CMAQ monies would be spent. The targets have been a useful tool in helping MPOs to obligate all or most of the CMAQ funds they are authorized to spend.

For the Boston MPO, CMAQ is viewed as a funding category, not a programmatic one. The MPO divides available funding for CMAQ projects into three programs: Suburban Mobility, TDM, and the Regional

Bike Parking Program, each with specific qualification guidelines and individual administrative committees or agencies. These programs receive approximately $1.5 million a year, with a set amount of funding for each, which leaves approximately $7 million of CMAQ target funding available for programming. The MPO distributes this money among projects selected from the larger TIP universe, which is the list of projects for which Federal funding has been requested. Selection of projects from this larger pool is done by the Transportation Planning and Programming Committee during the TIP development process. This is a bottom-up process: proposed projects work themselves up from the municipal to the regional level, where they must compete with other municipalities' projects, and they make the cut for TIP inclusion only if they are broadly beneficial.

CMAQ eligibility is determined by a State CMAQ Consultation Committee, which meets twice a year to determine if a project is eligible for CMAQ funding. MassDEP serves as a member on that committee. The State CMAQ consultation process is not technically a consultation or a prioritization of projects; it only provides a check for regional determinations on the eligibility of projects to qualify for CMAQ funds. After the eligibility determination, each MPO prioritizes and selects projects as part of the overall TIP project selection process described above.

Qualitative and Quantitative Measures

The Massachusetts EOT developed most of the methodologies the Boston MPO uses to quantify the estimated emissions changes and cost-effectiveness of proposed projects to determine their CMAQ eligibility. Currently, the Boston MPO uses EOT methodologies for the following project types:

- Bus replacement.
- New bus service.
- Bicycle and pedestrian projects.
- Park-and-ride facilities.
- Traffic flow improvements.

Not every CMAQ project undergoes the standardized quantitative analysis. The Boston MPO takes qualitative factors into account as well, to determine CMAQ eligibility when quantitative analysis is not possible or practical. All spreadsheets calculate emissions changes and cost-effectiveness for Year 1 of the project only. Focusing on the first year standardizes/normalizes calculations across project/modal types to create an even playing field and minimizes the number of assumptions built into the model. The State noted that there are continuing debates within the transportation and environmental agencies regarding how to measure and judge the short- versus long-term air quality impacts of different project types. While traffic signalization provides short-term benefits, the benefits diminish over the long term due to shifts in travel patterns. The fact that short-term benefits are easier to measure may explain why the eligibility process focuses on that area.

Reporting and Evaluation

The Boston MPO compiles information about its CMAQ projects and programs, which it forwards to EOT. EOT compiles all of the regional CMAQ programs along with the Statewide program and then forwards this report to FHWA for inclusion in the National CMAQ database.

For operations-type projects, EOT undertakes a variety of routine post-project analysis work. This includes collecting information such as daily ridership statistics on transit lines, monthly reports on shuttle bus ridership, and bicycle and pedestrian counts on paths and routes that have received Federal funds. These analyses can be used in combination with the CMAQ 3-year funding limit for operations-related projects to steer money away from unsuccessful projects toward projects and programs yielding better results. Part of the challenge with follow-up is to ensure a level playing field. The staff noted that given the limited planning resources, it is difficult to justify the additional cost to do post-project evaluation for all CMAQ projects, let alone non-CMAQ ones.

Denver, Colorado: Denver Regional Council of Governments (DRCOG)

Introduction

The Denver Regional Council of Governments (DRCOG) has received CMAQ funding since the program's creation in 1991. The size of the CMAQ pot has grown from $3 million annually under ISTEA to $18 million annually under SAFETEA-LU. This has allowed DRCOG to test new program areas and the program to evolve over time. The Denver MPO experienced frequent violations of NAAQS in the 1980s but has undertaken a number of measures to improve regional air quality since that time. As a result, the MPO is currently in maintenance for CO and PM 10. After several years in attainment, however, EPA redesignated most of DRCOG's planning area as "marginal" nonattainment for the 8-hour ozone standard in November 2007. The newly formed Denver/North Front Range nonattainment area covers nine counties in the Denver metropolitan region and spans portions of three separate regional transportation planning regions:

- DRCOG.
- NFRMPO, the Fort Collins MPO.
- Upper Front Range Transportation Planning Region (UFR TPR), a non-MPO transportation planning area covered by CDOT.

A Memorandum of Agreement was signed in March 2008, authorizing DRCOG and NFRMPO to determine conformity on behalf of the portion of UFR TPR that falls within the 8-hour ozone nonattainment area. Because the air quality boundaries are not the same as the planning region boundaries, DRCOG's transportation planning and conformity determination processes have become more complicated.

CMAQ Objectives and Procedures

The DRCOG committee structure, which includes the Transportation Advisory Committee (TAC), the Regional Transportation Committee (RTC), and a study session committee comprising DRCOG board members and the Metro Vision Issues Committee (MVIC), oversees transportation planning for the MPO and makes recommendations to the DRCOG board on funding of the TIP. The TIP process has established policy on how to fund different types of projects within its three discretionary Federal funding sources (CMAQ, Federal Surface Transportation Program [STP]-Metro, and STP-Enhancement). CMAQ is used to fund the following types of projects:

- TDM.
- ITS.
- TSSIP.
- RideArrangers program.
- FasTracks, the Denver region's 120-mile build-out of light and commuter rail, and FasTracks-related projects.
- Air quality improvement (e.g., alternative fuels, diesel retrofits).
- Bicycle and pedestrian projects.

When asked what would be different without the CMAQ program, DRCOG responded that, without the program but with the same overall funding authority, it would still have invested heavily in transit and bicycle/pedestrian projects but it might not have funded as many air-quality-focused projects.

Project Selection and Funding

DRCOG does not treat CMAQ as a stand-alone program; instead, CMAQ-type projects are solicited and selected as part of the 4-year TIP-update cycle. Eligible applicants include county and municipal governments; regional agencies, such as the Regional Transportation District (RTD), the Regional Air

Quality Council (RAQC), and DRCOG itself; and State agencies, such as CDOT and the CDPHE. Thus far, local governments and regional agencies have been most active in soliciting CMAQ funds through TIP. For the most recent TIP, DRCOG programmed $106 million in CMAQ projects for the FY 2008–2013 TIP, which is 4.3 percent of the $2.5 billion in the current TIP.

In making decisions about how to allocate its CMAQ funds, DRCOG's project selection criteria differ by project type. For each project type, criteria may be established that incentivize certain projects within those categories. Project types that are typically associated with CMAQ funding include air quality improvement, station area master plans, new bus service, and non-FasTracks transit passenger facilities. Bonus points are also available and depend on the project type, which can include diesel retrofits (a SAFETEA-LU priority) and overmatch of funds.

DRCOG's TIP application process is Web-based, with tailored evaluation criteria embedded in each project type. Eligible applicants submit project proposals online and are able to self-score their projects as they fill out the application. The scoring assessment is quantitative: applicants are asked to submit data relevant to the specific project-type category to which they are applying, using embedded formulas to generate final scores. DRCOG staff screen submitted applications for eligibility; they then review and correct the scores as necessary for consistency and quality control.

Qualitative and Quantitative Measures

Applicants are responsible for initially estimating the benefits of their proposed projects, using the evaluation criteria embedded in their Web application. Formulas embedded in each of DRCOG's project-type applications online generate quantitative estimates of proposed projects' benefits, such as emissions reduction, cost-effectiveness, and usage, for consideration in the prioritization and final selection process. In Phase I of project selection, 75 percent of DRCOG's CMAQ funding goes directly to the projects that scored highest within its 10 project category types, making quantitative analysis the primary focus. Target percentages for each funding category, such as CMAQ, are also preset for each project type to assist in project selection before the selection process begins. For example, for the first 75 percent of CMAQ funding, 70 percent will go toward air quality improvement projects. The maximum percentages awarded for the evaluation criteria in CMAQ-related-funding project types are as follows:

- 4 percent for diesel retrofit projects.
- 29 percent for projects that reduce VMT or directly reduce air pollutants.
- 29 percent for cost-effectiveness.
- 12 percent for additional local overmatch.
- 14 percent for supporting a Metro Vision strategic corridor (specific-project-related).
- 12 percent for Metro Vision implementation (sponsor-related).

In Phase II of project selection, projects for the remaining 25 percent of funding are selected by the DRCOG board by balancing the final evaluation scores with a series of concerns outside the project-type-specific criteria. Phase II priorities include:

- Financial equity of project awards across DRCOG members at the county level.
- Potential cost-savings from merging projects.
- Projects in DRCOG-defined strategic corridors.
- Project readiness for construction.
- Projects in very small communities (less than 10,000 population or employment).

Reporting/Post-Project Evaluation

When preparing information for reporting, DRCOG must gather information for two different reporting systems, FHWA's CMAQ database and CDOT's recently developed CMAQ Reporter. This creates administrative challenges since each system requires different information and calculations. In 2000, the Colorado Transportation Commission expressed concern about the effectiveness of the CMAQ program in improving air quality and adopted a resolution (TC-807) to increase accountability for how CMAQ funds are spent. This led to the development of the CMAQ Reporter, a database maintained by CDOT that requires fund recipients to report annually on the effectiveness of their CMAQ expenditures. MPOs are working with CDOT to develop methodologies in the CMAQ Reporter that will more accurately calculate future-year air quality benefits of various project types.

DRCOG conducts post-project evaluations on certain types of CMAQ projects but not all. Consultants are hired to prepare post-project evaluations, track the emissions reductions of diesel retrofit projects, and conduct follow-up surveys to see if people have changed their behavior as a result of one of DRCOG's outreach programs like carpool/vanpool or Bike-to-Work Day. For the latter, DRCOG conducts follow-up surveys to see if participation led to any other travel behavior changes and then tracks the Denver-area results relative to other cities organizing Bike-to-Work Days. This helps to track the project's progress from year to year and to compare it with that of other cities trying to support more bicycle transportation.

Fort Collins, Colorado: North Front Range MPO (NFRMPO)

Introduction

NFRMPO is the MPO for 15 local governments in northern Colorado. It covers an area of about 675 square miles with a population of about 440,000 residents. The MPO has been experiencing rapid growth; the population is projected to grow to 730,000 by 2035. The major urbanized areas in the MPO are Fort Collins and Greeley. The City of Fort Collins is in nonattainment for CO. This is the only area within NFRMPO that is eligible for CMAQ funding. The area meets NAAQS for nitrogen oxide and PM 10.

CMAQ Objectives and Procedures

Currently, CMAQ functions as a stand-alone program separate from the solicitation of other projects for TIP. Beginning with the next TIP update in 2010, NFRMPO will consolidate the call-for-projects process. At that time, CMAQ will be handled as part of the overall TIP funding process.

The NFRMPO CMAQ program consists of capital, transit, TDM, and service projects. Until 2005, the major focus within the MPO was on TDM measures. More recently, additional emphasis has been placed on intersection and traffic flow improvements. Because only the City of Fort Collins is eligible for CMAQ funding, the MPO must determine which portion of any proposed CMAQ project is contained within the city boundaries and where any benefits may accrue. For projects that are outside of the maintenance area, a graduated benefit amount is applied. For instance, a new transit service that operates both within and outside of the city boundaries can only claim emissions reductions for the portion of service that occurs within the city boundaries.

NFRMPO focuses its CMAQ funding efforts on air quality issues. Its CMAQ selection process does not directly address congestion issues. In 2007, the MPO adopted a Congestion Management Process (CMP), and there is a desire to better integrate the CMAQ program into the CMP.

The MPO requires that each project proposal be submitted by either a municipal or State agency. However, it does encourage other agencies and groups to cosponsor with a governmental agency. The MPO requires any local sponsor to provide a local match, usually 20 percent of the project cost. This local match can be either cash or in-kind services.

The 14-member CMAQ Evaluation Committee oversees MPO staff in the project evaluation and selection process. The committee consists of representatives from Federal agencies (FHWA, FTA, EPA), State agencies (CDOT, CDPHE), and the MPO.

Project Selection and Funding

Each sponsor must fill out a standard application including sections on purpose, project participants, scope of work, and evaluation process. In order to ensure that proposals meet the eligibility requirements, the project sponsors are requested to preview the scope with the MPO staff. The MPO staff has found that this can expedite the process of identifying potential problems early in the application process.

As part of its CMAQ Project Submittal Process Package available to all potential project sponsors, NFRMPO offers a word of caution about using Federal funds: "There are two caveats related to scale to bear in mind in pursuing a Federal-Aid project. First, the administrative burden of a Federal-Aid project is substantial. A very small project is often best accomplished with local funds to avoid the extra administrative burden. Second, the project scope and scale may expand because of Federal procedures."

Once projects have been submitted, the MPO uses a three-tiered scoring system to evaluate them, which includes:

1. Short-term air quality benefits (Year 1): 50 percent of total score.
2. Long-term air quality benefits (Years 2–5): 20 percent of score.

3. Regional planning achievement: 30 percent of score.

Bonus points are awarded for a local overmatch, multiagency or PPP projects, and multimodal projects. Specific preference is not given to diesel retrofit proposals, but historically these types of projects have scored well and have received funding.

The scoring of projects is done by the staff, with review by the CMAQ Evaluation Committee. The Evaluation Committee then compiles a list of projects for proposed inclusion in TIP. This list might include suggestions for the scopes of projects to be modified or staged over multiple years. The list is then approved by the MPO board for inclusion in the regional TIP.

Qualitative and Quantitative Measures

In 2004, NFRMPO moved from a qualitative project selection process to one that relies heavily on a quantitative process to determine projects that will receive CMAQ funding. This process was introduced to help balance the competition for funding and to remove some of the perceived political influence. To help standardize the quantitative analysis, NFRMPO hired a consulting firm to oversee the development of calculations for CMAQ programs and projects.

While the responsibility for estimating air quality benefits rests with project sponsors, MPO staff and the air quality consultant review their assumptions, and at times request that they recalculate the air quality benefits analysis post-review. The air quality consultant supports NFRMPO staff with the technical expertise to ensure that the assumptions and estimates are reasonable.

Reporting/Post-Project Evaluation

As part of the CMAQ proposal submittal process, project sponsors must submit an evaluation plan. This is usually a brief statement of how the sponsor will measure the success of the program. This information is then shared with NFRMPO. Part III of the application states: "Projects should have a detailed evaluation process identified prior to application submittal. This process should provide means of evaluating the effectiveness of the project over time, its ability to meet project goals and objectives, and quantify the air quality benefits." Beginning in 2009, CDOT now requires this evaluation plan as part of the contract for all of its CMAQ-funded projects.

Due to Colorado State requirements, NFRMPO must collect data for two reporting systems: the FHWA CMAQ database and the Colorado Transportation Commission's Colorado Reporter, the State's CMAQ reporting system. For the FHWA CMAQ database, NFRMPO provides the data and information to CDOT. CDOT then compiles the information for the State and transmits the data to FHWA. The Colorado Reporter requires additional data from each CMAQ project. While FHWA requests data on the first year of operation of a CMAQ project, Colorado requires a life-cycle estimate of benefits.

Medford, Oregon: Rogue Valley Council of Governments (RVCOG)

Introduction

RVCOG is one of six MPOs in Oregon. Located in southern Oregon, RVCOG has a 22-member board of directors, including 15 local governments and seven special districts such as higher education, economic development, soil and water conservation, and regional transit. The RVCOG board delegated responsibility for the MPO to the Rogue Valley Metropolitan Planning Organization (RVMPO) Policy Committee. The Policy Committee comprises eight local member jurisdictions, an ODOT representative, and a representative from the regional transportation district. RVCOG staffs RVMPO. The Rogue Valley was first designated as nonattainment for PM 10 in 1990 but moved into maintenance in 1996. The Medford Urban Growth Boundary is also designated "moderate" maintenance for CO. Both maintenance areas fall within RVCOG's transportation planning region.

There is no Statewide CMAQ program in Oregon. ODOT uses its own formulas to distribute CMAQ funds among the nonattainment and attainment areas rather than relying on Federal apportionment formulas. In the past, RVMPO received only about $400,000 per year in CMAQ funds, and member jurisdictions divided the funds among themselves in a relatively informal process. ODOT recently adjusted its CMAQ allocation formulas, however, which more than doubled the size of RVMPO's allocation to more than $1 million in annual CMAQ funding. This increase spurred RVMPO to develop more formal, standardized processes for distributing CMAQ funds within the region. For RVMPO's 2008–2011 TIP, CMAQ projects accounted for about 2.5 percent of the total funds programmed ($4.8 million out of a $198 million TIP).

CMAQ Objectives and Procedures

RVMPO treats CMAQ as a stand-alone program, with its own solicitation, public input, and selection processes. CMAQ is a critical component of RVMPO's transportation program as it accounts for 70 percent of the discretionary funds over which RVMPO has direct programming authority. CMAQ is also an integral component of RVCOG's air quality conformity process. Emissions reduction credits from CMAQ projects are applied during conformity determinations and are the MPO's sole mechanism for mitigating onroad mobile source emissions. Staff noted that without these "CMAQ credits," the MPO's ability to meet conformity requirements would be in question.

The primary focus of RVMPO's CMAQ program is addressing the MPO's PM and air quality problems. The agency's official CMAQ solicitation packet states four objectives for CMAQ-funded projects:

1. Enable the MPO to maintain the NAAQS.
2. Meet regional air quality and transportation needs.
3. Meet multimodal objectives.
4. Meet State and local goals and objectives (e.g., reduce reliance on automobiles).

As the size of the CMAQ program has grown, the RVMPO Policy Committee has formalized the decisionmaking process to ensure a fair distribution of funds based on project quality. The board developed and adopted a single set of quantitative criteria that it now uses to evaluate all CMAQ project proposals. The RVMPO Policy Committee worked closely with member jurisdictions to develop these criteria, trying to build an evaluation tool that would be flexible over time and allow the program to fund new and innovative types of projects.[23] The new process emphasizes the notion that CMAQ monies are awarded to projects, not jurisdictions. RVCOG staff report that moving toward a more formal process like this helped to improve the efficiency, fairness, and effectiveness of the program.

---

[23] For this reason, diesel retrofits was added as one of the evaluation criteria, though at a lower point value than CO/PM 10 reductions in order to exert influence without skewing project rankings.

Project Selection and Funding

RVCOG staff issue a call for CMAQ projects every 2 years in conjunction with the TIP update cycle. The solicitation is an open application process that lasts for 2 months. Public input is conducted through public notices, public comment during RVMPO committee meetings, and review and recommendations on project funding by a Public Advisory Council (PAC) comprising 11 residents from member municipalities. RVMPO typically uses CMAQ to fund short-term projects (i.e., those that can be completed within one TIP cycle).[24]

Public, private, and nonprofit agencies are eligible to apply, but local municipalities have been the most common and successful applicants for CMAQ projects in the past. The Oregon DEQ has submitted applications as well, such as for the first diesel retrofit funded in the region. A new focus on PPPs and diesel retrofits means that private businesses also are increasingly submitting applications.

Once solicitation is complete, 1 month is set aside for project grading and evaluation. RVCOG staff present applications and work with members of the TAC to evaluate projects and develop rankings. The TAC comprises 16 municipal representatives (city managers and a mix of planning, public works, and economic development staff) and six representatives from other regional and State agencies (regional transit district, ODOT, Oregon DEQ, and Oregon Department of Land Conservation and Development [DLCD]). A single set of evaluation criteria is used to rank and score all projects. Projects can earn up to 115 points across 10 criteria such as CO/PM 10 reduction, cost-effectiveness, and likelihood of reducing reliance on automobiles. Once the TAC has evaluated all project proposals and tallied all scores, it makes recommendations to the RVMPO Policy Committee, which has final discretion over ultimate funding. The Policy Committee comprises eight local member jurisdictions, an ODOT representative, and a representative from the regional transit district.

Qualitative and Quantitative Measures

RVMPO uses standardized calculations to estimate the specific air quality benefits of each proposed project prior to that project's overall evaluation and ranking. It also uses standardized calculations to estimate the VMT reduction benefits of several types of projects:

- Transit.
- Bikes on Buses.
- Employer Trip Reduction Program.
- Employer Group Bus Pass Program.
- Park-and-Ride.
- Rideshare.

Sponsors are asked to provide the information needed to perform these calculations and are given guidance to explain how emissions benefits will be derived to increase the transparency of the process. They are not asked to complete the calculations themselves; this is done by MPO staff. Currently, emissions benefits are calculated only for CO and PM 10. This may change, as the State is now strongly encouraging MPOs to address carbon emissions and climate change.

Reporting/Post-Project Evaluation

Using information provided by the MPOs, ODOT staff are responsible for reporting to FHWA's CMAQ database. Post-project evaluation is required by ODOT for all diesel retrofit projects. RVCOG staff noted the benefits of conducting post-project evaluation on all project types but explained that it is not currently possible due to limited staffing and resources.

---

[24] Because of this, CMAQ projects are rarely part of the Long-Range Transportation Plan.

RVMPO has found that, due to the reporting, administrative, and design requirements that come with the use of Federal funds, it is not efficient to use Federal funds for projects costing less than $200,000. Local jurisdictions estimate that complying with the requirements of using Federal money increases their project costs by 60 percent, and they are trying to work out an arrangement with the State to swap CMAQ funds for State funds that would not have the same requirements.

Pittsburgh, Pennsylvania: Southwestern Pennsylvania Commission (SPC)

Introduction

SPC is one of 15 MPOs in Pennsylvania. SPC's planning region covers 10 counties and is served by 10 fixed-route transit agencies. The governing board includes representatives from the City of Pittsburgh, the region's counties and transit agencies, and State and Federal agencies such as PennDOT, the State economic development agency, FHWA, and FTA. The SPC planning region comprises nine separate NAAs and maintenance areas within southwestern Pennsylvania for five separate air quality standards. These include the Pittsburgh Central Business District for CO, one area within Allegheny Country for PM 10, three separate areas for the 8-hour ozone standard, three separate areas for the annual PM 2.5 standard, and one area for the daily PM 2.5 standard.

SPC recently refined its CMAQ project selection process in response to new provisions related to the enactment of SAFETEA-LU as well as to formalize what had previously been a more informal, ad-hoc process. As part of this reexamination, a consultant was hired to facilitate the update of SPC's CMAQ project solicitation, evaluation, prioritization, and selection procedures. SPC identified examples of MPOs around the country with noteworthy CMAQ practices to guide its update and also used FHWA's CMAQ guidance and integrated suggestions from the Transportation Research Board's (TRB) Special Report 264 – Congestion Migitation and Air Quality Improvement Program.[25]

SPC solicits and programs CMAQ projects in conjunction with the TIP update cycle: a four-year TIP is updated every 2 years. The SPC region receives about $25 million a year in CMAQ funding, which accounts for about 3 percent of total TIP funding. SPC has done focused outreach to public agencies and the public to bring them into the process and have a wider range of participants.

PennDOT suballocates all CMAQ funding directly to the MPOs; there is no separate Statewide CMAQ program at the State level. If PennDOT has a project it would like to fund using the CMAQ program, the appropriate PennDOT District Office must submit a project proposal to the MPO for funding consideration, like any other project sponsor. Although PennDOT does not maintain its own stand-alone CMAQ program, it does take a leadership role in supporting and coordinating MPO and CMAQ efforts. At the beginning of each TIP update cycle, PennDOT and MPO staff from around the State meet to discuss strategies and techniques for preparing individual TIPs and evaluating project proposals within Federal and State transportation programs. The result of this effort is concurrence on financial and procedural guidance for the TIP update. This guidance document establishes fiscal targets for programs, including the allocation of CMAQ funding for each of the qualifying MPOs.

Objectives and Procedures

The SPC CMAQ process, as endorsed by the board, has established priority for four types of projects for CMAQ funds: signalization, diesel retrofits, TDM, and bicycle/pedestrian. CMAQ projects are solicited in conjunction with the TIP update cycle. As such, there is no stand-alone public input process for CMAQ projects.

Proposed projects that are not selected in one round may be resubmitted and may recompete for funding during the next TIP update and CMAQ solicitation. In order to do so, they must update costs, assumptions, emissions benefits calculations, and scopes (as appropriate). Under SPC's revised CMAQ process, all proposed CMAQ projects, both new and resubmitted, are subject to quantitative analysis and are then ranked on current costs, assumptions, and estimated costs and benefits during each update to the TIP.

---

[25] The publication can be found at the following link: http://www.nap.edu/catalog.php?record_id=10350

As SPC has opened its CMAQ program processes and worked to expand its reach, it has seen a rise in the number of "nontraditional" transportation project applications such as those for diesel retrofits and car-share programs. Bringing more nontraditional sponsors into a Federal funding process can lead to new types of projects, but it also has challenges, especially in terms of educating project sponsors about Federal requirements for activities such as contracting and financing.

Project Selection and Funding

SPC has a well-defined evaluation process. Staff color-code each CMAQ project proposal by type/category and then rank projects within those categories to facilitate fair and logical comparisons. In the past, review of applications was done by staff. Now, all applications are packaged and distributed on a CD to the committee members for their review and evaluation to identify unrealistic assumptions or benefits. After ranking each proposal within project categories, the CEC develops a recommended list of projects for TIP. The selections are made by the CEC through secret ballot.

Without the CMAQ program, SPC staff felt that they would not be able to fund some current innovative projects like the City's car-sharing program with the Pittsburgh Downtown Partnership (PDP) or marine diesel retrofits with the Port of Pittsburgh Commission. These projects confer quality-of-life and air quality benefits to the residents of southwestern Pennsylvania but may not have traditionally been thought of as "transportation" projects or may not be able to be funded through other federal transportation funding programs.

Qualitative and Quantitative Measures

Projects are ranked using three separate measures. First, SPC quantifies the air quality benefits for each project proposed for CMAQ funding. All projects are evaluated for five air quality and cost/benefit factors, using a standardized set of analysis models developed by PennDOT. One of the factors measures cost-effectiveness in reducing air pollution by calculating the cost-per-unit change in emissions. Second, projects that are within one of the four board priority categories are given preference. Third, projects are rated according to a set of nine ancillary factors that include:

- Overmatch.
- Raising of public awareness of TDM options.
- Consistency with the region's long-range plan.
- Whether they bring nontraditional funding to TIP.
- Congested corridor rating.

PennDOT maintains an on-call consultant for Statewide air quality issues and analysis. These consultant services are made available to the State's MPOs to assist in quantifying the benefits of new projects such as the City's car-sharing proposal, which arise in new funding rounds and are not easily quantified by traditional standardized analysis tools. Staff estimate that this accounts for about 10 percent of new projects. SPC staff expressed that, from an MPO perspective, having this kind of State leadership and support on technical assistance was an invaluable component in improving their own process.

Reporting and Evaluation

PennDOT has developed a report card to provide information on how money is being allocated to MPOs around the State and other financial management tools. This report card is prepared on a quarterly basis and strives to provide the most up-to-date information on obligations that have been made.

SPC conducts post-project evaluation of some projects to examine actual versus predicted benefits. Staff conduct before-and-after studies for some traffic signalization projects—for example, to demonstrate whether a completed project achieved the benefits projected in the CMAQ project evaluation process. SPC staff noted that as post-project evaluation becomes more routine for traffic signalization projects, sponsors of other types of projects may feel compelled to conduct post-project evaluation to be able to demonstrate cost-effectiveness, congestion mitigation, or air quality benefits. Thus, a trend toward having project sponsors build post-project evaluation into their project plans and budgets may be emerging in subsequent CMAQ funding rounds.

San Francisco, California: Metropolitan Transportation Commission (MTC)

Introduction

MTC is one of 19 MPOs in California. It serves as the nine-county San Francisco Bay Area MPO and includes such major cities as San Francisco, San Jose, and Oakland. Its 19-member policy board comprises primarily local elected officials (mayors and county commissioners), along with one representative each from the Association of Bay Area Governments and the Bay Area Conservation and Development Commission. Three nonvoting members also participate on the board to represent important State and Federal interests, such as Caltrans, FHWA, and the Department of Housing and Urban Development (HUD).

MTC's planning region overlaps with three air basins (San Francisco, Sacramento, and Northern Sonoma), two of which are in nonattainment:

- San Francisco Bay Area basin: "marginal" for 8-hour ozone, maintenance for CO.
- Sacramento basin: "serious" nonattainment for 8-hour ozone, maintenance for CO.

Caltrans distributes CMAQ funds to MPOs around the State on the basis of population and the severity of air quality. Caltrans does not run a Statewide CMAQ program per se, but it does reserve a small portion of funds for the five non-MPO areas around the State that qualify to receive CMAQ dollars. CMAQ accounts for approximately 2.3 percent ($300 million) of MTC's current (2008–2013) $13 billion TIP.

CMAQ Objectives and Procedures

As in other MPOs, programming of CMAQ projects takes place in conjunction with the TIP update process. CMAQ is not treated as a stand-alone program in the San Francisco Bay region but as an integrated part of the long-range transportation planning process. MTC uses planning goals and measures of effectiveness outlined in the long-range plan to develop comprehensive and multimodal program areas, each with its own specific goals and objectives. Program areas include:

- Transportation for Livable Communities/Housing Incentive.
- Regional Bicycle and Pedestrian.
- Lifeline Program, Free Transit.
- TransLink® (universal fare card).
- Regional Rideshare.
- Traffic Operations System/Incident Management.
- Clean Air Initiatives.

MTC pools the MPOs' CMAQ funds with Federal STP funds and other State and local funding sources to support projects within each of these program areas. Only at the end of the planning and programming process are specific funding sources matched up with individual selected projects. This means that the project sponsors must be aware of the program area to which they are applying but not the funding source that will cover their projects.

Worsening air quality and traffic congestion are two of the major challenges facing the San Francisco Bay MPO, so MTC routinely has many more CMAQ-eligible projects submitted for funding than it has available CMAQ dollars to spend. Still, CMAQ is seen by MTC as a means of further supporting its air quality and congestion goals as well as projects with quality-of-life benefits that may not be eligible for other funding sources. Any public agency with a role in surface transportation planning and implementation may submit projects for consideration by MTC's program committees.

Project Selection and Funding

MTC staff use FHWA guidance to establish which projects are eligible to receive CMAQ funds, but they do not have one standard approach for selecting the projects that will ultimately receive the funds. Each of MTC's multimodal program areas runs its own project solicitation, public outreach, and selection process. Evaluation and selection criteria are tailored to the specific needs of each program to which a CMAQ-eligible project may apply, and funding sources are matched to individual projects only during the project selection process.

Each MTC program area uses different rules and procedures for project evaluation; different committees and member agencies take the lead in selecting projects, depending on the program. With the Transportation for Livable Communities program, for example, two-thirds of the available funding is chosen at the regional level by the MTC board, but one-third is split by formula and distributed directly to member counties, where local agencies and officials determine which projects to fund. In the Regional Bicycle and Pedestrian Program, however, only a quarter of all project selection decisions are made by the MTC board. The other 75 percent of funds are suballocated directly to County Congestion Management Agencies for project selection, subject to MTC program rules and criteria.

MTC staff believe that many of their current projects and programs would continue to be funded even if the CMAQ program ceased to exist. This is because the main criterion for funding projects across all programs and funding categories is consistency with the vision of the long-range plan in which air quality improvement and reduced congestion are primary goals.

Qualitative and Quantitative Measures

As with project selection, the calculation of project benefits depends on the program area within which projects are being proposed. MTC uses quantitative methods to evaluate many of its projects, but qualitative indicators are also used to evaluate proposals having benefits that are difficult to calculate with precision due to the size or nature of the project. Marketing projects are evaluated qualitatively, though quantitative analysis may be used as supporting evidence; for example, a follow-up survey may be conducted to ascertain how many people heard a given radio spot or saw a newspaper advertisement. The decision as to which qualitative measures to use and when to use them is left to the discretion of each of MTC's individual program areas.

In terms of quantitative analysis, MTC staff use a number of methodologies to help estimate the benefits of CMAQ-eligible projects. MTC relies heavily on calculations outlined in *TRB Special Report 264* on the CMAQ program, and on a report by the California Air Resources Board (CARB), to evaluate many of the projects submitted for consideration. However, MTC has developed several calculation methods in-house for new or nontraditional project proposals that do not clearly match the guidance outlined in those two resources.

CARB and BAAQMD staff provide additional expertise. MTC staff work very closely with BAAQMD staff to develop air quality benefit calculations for new and nontraditional projects and rely on them for expert guidance on which project proposals will have the best impacts.

MTC staff also work closely with project sponsors to estimate project benefits once specific methodologies are developed or adopted since many of the calculations are too complex for sponsors to conduct on their own. In response to recently enacted statewide climate-change legislation, MTC and other MPOs in California are developing methods to estimate the carbon dioxide (CO_2) impacts of projects and will begin including those benefits calculations in project evaluations moving forward.

Reporting and Evaluation

MTC staff provide Caltrans with the necessary project information for reporting to FHWA's CMAQ database, but Caltrans retains responsibility for managing this process and inputting the final information.

Staff at the State and MPO levels expressed frustration with the administrative challenge of complying with Federal reporting requirements.

MTC does not have an across-the-board policy to measure the impacts and benefits of implemented projects, but it does conduct limited post-project evaluations, particularly to ascertain and document the cost-effectiveness of various project types. An evaluation of the region's FSP/Incident Management program revealed it to be particularly cost-effective, which provided the basis for MTC's ongoing funding to support and expand the program. Now, staff are developing cost-effectiveness tools to assess the benefits of various TCMs as well as the region's Free Transit Program.

Appendix A: CMAQ Phase II Site-Visit Locations and MPO Points of Contact

Site-Visit Locations

Location	U.S. Region	MPO Size*	FHWA Division Office (Resource Center)	FTA Regional Office	EPA Regional Office
Boston, MA	Northeast	XL	Boston, MA	Boston, MA – Region 1	Boston, MA – Region 1
Pittsburgh, PA	Midwest	L	Harrisburg, PA	Philadelphia, PA – Region 3	Philadelphia, PA – Region 3
Birmingham, AL	South	L	Montgomery, AL	Atlanta, GA – Region 4	Atlanta, GA – Region 4
Jackson County, OR (Rogue Valley)	West	S	Salem, OR	Seattle, WA – Region 10	Seattle, WA – Region 10
San Francisco, CA	West	XL	Sacramento, CA (San Francisco)	San Francisco, CA – Region 9	San Francisco, CA – Region 9
Denver, CO	Mountain	L	Lakewood, CO (Lakewood)	Lakewood, CO – Region 8	Denver, CO – Region 8
Fort Collins, CO	Mountain	M	Lakewood, CO (Lakewood)	Lakewood, CO –- Region 8	Denver, CO – Region 8

*XL = extra-large, L = large, S = small, and M = medium.

MPO Points of Contact

Location	MPO Contact	Telephone	E-mail
Boston MPO/ Mass. EOT	Pam Wolfe	617-973-7141	pwolfe@ctps.org
Pittsburgh, PA	Chuck Imbrogno	412-391-5590, x319	Imbrogno@spcregion.org
Birmingham, AL	Bill Foisy	205-251-8139	bfoisy@rpcgb.org
Jackson County, OR (Rogue Valley)	Vicky Guarino	514-423-1361	vguarino@rvcog.org
San Francisco, CA	Craig Goldblatt	510-817-5837	cgoldblatt@mtc.ca.gov
Denver, CO	Todd Cottrell	303-480-6737	tcottrell@drcog.org
Fort Collins, CO	Tia Raamot	970-224-6102	traamot@nfrmpo.org

APPENDIX B: CMAQ PHASE II SITE VISITS—INTERVIEW GUIDE

USDOT and EPA are conducting a series of site visits this summer to better understand the CMAQ project selection and implementation practices and to document the most effective approaches. This effort will result in a report, as required under SAFETEA-LU, evaluating the CMAQ program.

The following questions are intended as a guide for our series of interviews with your designated State and regional transportation and air quality experts. Federal staff from FHWA, FTA, and EPA will be present for the interviews. The interviews focus on broad questions, with specific questions available as prompts for gathering information about best practices.

Introductory/Background Questions

1. Please provide contextual information on the nonattainment areas, including:
    a. Population and growth trends.
    b. Travel trends (e.g., VMT growth) and congestion problems (e.g., significance of congestion problems faced).
    c. Type of nonattainment or maintenance area status, and changes in status.

2. Provide an overview of the organizations involved in funding, planning, implementing and evaluating CMAQ-funded projects. What role does each of these organizations play in the process?

CMAQ Program Objectives

3. How does the CMAQ program fit into local transportation plans and objectives?
    a. If CMAQ program funding were not available, would similar CMAQ-like projects be undertaken, and what funding sources would be used?
    b. Are there particular types of projects that would not likely be funded without the CMAQ program?

4. What do you see as the primary goal of your CMAQ program? Are there other objectives addressed by the program that you see as important (e.g., mobility enhancement, community livability)? Please describe.

5. What role does the CMAQ program play in your area's air quality planning process and conformity requirements for meeting regional air quality goals and/or National standards?

6. Which types of CMAQ projects have you found to be, or do you believe are, most effective in reducing emissions in the short term and long term? Does your selection process focus on short-term or long-term benefits? Why?

7. Have PPPs or other innovative financing techniques been used to fund CMAQ activities?

Local CMAQ Program Procedures

8. Please describe how projects are **initiated** as candidates for CMAQ funding.
    a. Which agency has the primary responsibility for identifying or soliciting proposals for CMAQ funding? Are other agencies involved?

b. Is there a structured process for nominating candidates for CMAQ funding? If so, please provide relevant application forms, calls for projects, or other relevant documentation. What are the required elements in the project proposal or application?
c. What agencies/organizations are eligible to propose projects? Which of these have been most active in proposing projects?
d. Are any project types encouraged (e.g., diesel retrofits, PPPs)?
e. Is there a regular cycle for new project proposals annually, or on a less frequent basis (e.g., in association with updates to the metropolitan transportation plan or TIP)?

9. Please describe how project air quality benefits are **analyzed.**

   a. Which agency or agencies calculates the emissions benefits (e.g., project sponsor, MPO, State DOT)?
   b. Is guidance on calculating emissions benefits provided to those generating proposals? If so, by whom? What is the nature of this guidance?
   c. To what extent is the State air agency (or regional air quality district) involved in the process?
   d. What techniques are used to estimate travel effects and emission reductions for CMAQ projects? Are models and modeling techniques used? Is this true for all project categories? If not, what other methods are being used? Please describe.
   e. Are National, regional, or local values typically used for calculation of such factors as trip length and average vehicle occupancy?
   f. For what pollutants are emissions benefits calculated (e.g., CO, NO_x, VOC, PM 10, PM 2.5, CO_2, other)?
   g. Are post-project evaluations conducted for certain types of projects? In which cases? Who collects these data?
   h. Are the air quality impact estimates reviewed as part of the project selection process? Have changes been required or made in project design or selection as a result of air quality evaluations?
   i. Do you have any comments on the quality of analyses? What are the key challenges you see? Have actions been taken to improve the quality of analyses?
   j. Have you made any improvements over the past 5 years in the analysis of CMAQ funds that you would consider a "best practice"?

10. Please describe how projects are **selected** for CMAQ funding.

    a. Is the funding for a CMAQ project done separately or as part of the overall TIP funding process?
    b. Is there a formal project selection process? If so, please describe and provide documentation.
    c. To what extent is public input obtained? How is the process transparent to an interested citizen or agency?
    d. Is the procedure largely quantitative in nature (e.g., project rankings, scoring across different criteria) or qualitative?
    e. Who is involved in the selection process (e.g., State DOT, MPO, air agency, EPA, FHWA)? Is there a formal selection committee structure? To what extent are air quality agencies involved?

f. To what extent are air quality benefits and contribution toward meeting conformity considered in project selection?
g. To what extent are potential congestion benefits considered in project selection? Are these effects measured, and if so, by which agency?
h. To what extent is cost-effectiveness considered in project selection? How is cost-effectiveness calculated? What have been your findings regarding project cost-effectiveness?
i. Have there been any opportunities to pursue innovative or less traditional projects and programs?
j. To what extent is priority given to diesel retrofits?
k. To what extent are other project outcomes (e.g., effects on greenhouse gases, economic development, social equity, community livability) considered in project selection? Are these effects related to goals in the metropolitan transportation plan? Are the effects reported or measured? If so, by which agency?
l. Have you made any improvements over the past few years in the selection process? If so, what are they?

11. Funding of CMAQ.

    a. Describe the capital programming process for CMAQ projects. How is the decision made to use CMAQ funds versus other Federal or State funds for a specific project?
    b. Are CMAQ projects ever funded using other Federal funding categories or innovative finance tools?
    c. What has been your percentage of programming of CMAQ funds for the past 3 years?
    d. What factors may have affected how much CMAQ funding was programmed?
    e. Do you have any best practices to share in terms of how you program Federal funds including CMAQ?
    f. To what extent is local match considered a factor in funding?
    g. How do you balance the competing needs for funding?

12. Which agency is responsible for **reporting** CMAQ project data to FHWA?

    a. What information, if any, is gathered in addition to the reporting data required by FHWA?
    b. In your opinion, is there additional information that should be gathered or reported to FHWA?
    c. Would you recommend any changes in the FHWA reporting process?

13. Are ex-post-project evaluations (evaluations after the project is complete) undertaken to determine whether desired travel changes, emissions reductions, and other project outcomes have been achieved? (Please provide copies of any such studies or analyses, if available.) If so, how is this information used by local or State agencies? If not, why are such evaluations not undertaken?

CMAQ Program Evaluation

14. What do you see as the main strengths and weaknesses of the CMAQ program as it is handled in your region?

15. What effects, if any, has the program had on agency or interagency decisionmaking at the State or regional level?

16. If there are things you could change about how the CMAQ program operates in your region, what would they be?

Specific Program/Project Best Practices

17. Discuss two specific CMAQ projects funded within the past 5 years, as well as what you think may be a best practice for the Nation.

    a. What distinguished this project?
    b. Describe the funding for the project.
    c. How will you track this project as it moves from being programmed on the TIP to implementation?

APPENDIX C: ACRONYMS AND ABBREVIATIONS

ADEM	Alabama Department of Environmental Management
ALDOT	Alabama Department of Transportation
APCA	Alabama Partners for Clean Air
AQ	air quality
BAAQMD	Bay Area Air Quality Management District
BFC	Bike Fort Collins
BJCTA	Birmingham-Jefferson County Transit Authority
CAA	Clean Air Act
Caltrans	California Department of Transportation
CARB	California Air Resources Board
CDOT	Colorado Department of Transportation
CDPHE	Colorado Department of Public Health and Environment
CEC	CMAQ Evaluation Committee (Pittsburgh)
CCF	crankcase filter
CHP	California Highway Patrol
CMAQ	Congestion Mitigation and Air Quality
CMP	Congestion Management Process
CO	carbon monoxide
CO_2	carbon dioxide
DEP	Department of Environmental Protection
DEQ	Department of Environmental Quality (Oregon)
DLCD	Department of Land Conservation and Development
DOC	diesel oxidation catalysts
DOT	Department of Transportation
DRCOG	Denver Regional Council of Governments
EOEEA	Executive Office of Energy and Environmental Affairs (Massachusetts)
EOT	Executive Office of Transportation (Massachusetts DOT)
EPA	Environmental Protection Agency
FSP	Freeway Service Patrol
FHWA	Federal Highway Administration
FTA	Federal Transit Administration
FY	Fiscal Year
HC	hydrocarbons
HUD	Department of Housing and Urban Development
I&M	Inspection and Maintenance
ISTEA	Intermodal Surface Transportation Efficiency Act
ITS	Intelligent Transportation Systems
MassDEP	Massachusetts Department of Environmental Protection
MPO	Metropolitan Planning Organization
MTC	Metropolitan Transportation Commission (San Francisco Bay Area)
MVIC	Metro Vision Issues Committee
NAA	nonattainment area
NAAQS	National Ambient Air Quality Standards
NFRMPO	North Front Range MPO (Fort Collins, Colorado)
NO_X	nitrogen oxide
ODEQ	Oregon Department of Environmental Quality
ODOT	Oregon Department of Transportation

PAC	Public Advisory Council
PDP	Pittsburgh Downtown Partnership
PennDOT	Pennsylvania Department of Transportation
PM 2.5	particulate matter (fine)
PM 10	particulate matter
PPM	parts per million
PPP	public-private partnerships
RAQC	Regional Air Quality Council (Colorado)
RPCGB	Regional Planning Commission of Greater Birmingham
RTC	Regional Transportation Committee
RTD	Regional Transportation District
RTP	Regional Transportation Plan
RVCOG	Rogue Valley Council of Governments (Medford, Oregon)
RVMPO	Rogue Valley MPO (Medford, Oregon)
SAFE	Service Authority for Freeways and Expressways (MTC)
SAFETEA-LU	Safe, Accountable, Flexible, Efficient Transportation Equity Act: A Legacy for Users
SFBCDC	San Francisco Bay Conservation and Development Commission
SIP	State Implementation Plan
SPC	Southwestern Pennsylvania Commission (Pittsburgh)
STIP	State Transportation Improvement Program
STP	Surface Transportation Program
TAC	Technical Advisory Committee, Transportation Advisory Committee
TCM	transportation control measures
TDM	transportation demand management
TE	Federal Transportation Enhancements Funding
TEA-21	Transportation Equity Act for the 21st Century
TIP	Transportation Improvement Program
TMA	Transportation Management Area
TMA	Transportation Management Association
TOD	time of day
TRB	Transportation Research Board
TSSIP	Traffic Signal System Improvement Program
UFR TPR	Upper Front Range Transportation Planning Region
USDOT	U.S. Department of Transportation
VMT	vehicle-miles traveled
VOC	volatile organic compound

www.ingramcontent.com/pod-product-compliance
Lightning Source LLC
Chambersburg PA
CBHW081845170526
45167CB00007B/2910